BestMasters

Springer awards „BestMasters" to the best master's theses which have been completed at renowned universities in Germany, Austria, and Switzerland.

The studies received highest marks and were recommended for publication by supervisors. They address current issues from various fields of research in natural sciences, psychology, technology, and economics.

The series addresses practitioners as well as scientists and, in particular, offers guidance for early stage researchers.

Karl Striegler

Modified Graphitic Carbon Nitrides for Photocatalytic Hydrogen Evolution from Water

Copolymers, Sensitizers and Nanoparticles

 Springer Spektrum

Karl Striegler
Leipzig, Germany

BestMasters
ISBN 978-3-658-09739-4   ISBN 978-3-658-09740-0 (eBook)
DOI 10.1007/978-3-658-09740-0

Library of Congress Control Number: 2015937602

Springer Spektrum

Printed on acid-free paper

Springer Spektrum is a brand of Springer Fachmedien Wiesbaden
Springer Fachmedien Wiesbaden is part of Springer Science+Business Media
(www.springer.com)

Geleitwort

Die photokatalytische Herstellung von Wasserstoff als einem viel versprechenden Energieträger ist eine der wichtigsten Herausforderungen der derzeitigen Forschung im Bereich der heterogenen Katalyse. Insbesondere sind noch wenige Katalysatoren bekannt, mit denen sichtbares Licht für die Wasserspaltung genutzt werden kann. Unter den zahlreichen Katalysatoren stellen graphitische Kohlenstoffnitride (g-C_3N_4) attraktive Kandidaten dar. Bisher gelingt die Wasserstoffbildung jedoch nur mit geringen Geschwindigkeiten oder unter Verwendung von Opferreagentien und erfordert häufig energiereiche UV-Strahlung. In seiner Masterarbeit verfolgt Herr Karl Striegler vor diesem Hintergrund drei Ansätze der Modifizierung, um aktivere und stabilere Katalysatoren auf der Basis von Kohlenstoffnitriden für die Wasserstoffbildung mit sichtbarem Licht zu erhalten: (i) das Einbringen von S- und O-haltigen Funktionalitäten, (ii) die Sensibilisierung durch Farbstoffe und (iii) die Immobilisierung von Edelmetall- und Cobaltoxid-Nanopartikeln. Die photokatalytischen Tests zur Wasserstoffbildung führt Herr Striegler in Anwesenheit von Methanol als Co-Reduktionsmittel durch. Die Materialien charakterisiert er umfangreich und führt auch DFT-Berechnungen von Bandlücken für die modifizierten Materialien durch. Herr Striegler wählt damit insgesamt einen sehr breiten Ansatz in einem noch jungen, sehr aktuellen und anspruchsvollen Gebiet der Katalyseforschung. Herr Striegler hat eindrucksvolle Ergebnisse erzielt. So konnten durch Copolymerisation von Dicycandiamid und Melamin Materialien mit höherer Wasserstoffbildungsgeschwindigkeit erhalten werden als bisher in der Literatur berichtet. S- und O-haltige Kohlenstoffnitride konnten hergestellt werden, mit denen eine Steigerung der Wasserstoffbildungsgeschwindigkeit um den Faktor 1.5 erzielt werden konnte. Durch kovalente Anbindung des Farbstoffs Erythrosin B konnte im sichtbaren Bereich eine höhere Aktivität als nur durch Imprägnierung erzielt werden. Für die Modifikation mit Pt- und Rh-Nanopartikeln (von enger Partikelgrößenverteilung) hat Herr Striegler eine neue Syntheseroute ohne Notwendigkeit zur Nutzung von UV-Strahlung erarbeitet. Für mit Co_3O_4-Partikeln modifizierte Kohlenstoffnitride konnte er erstmals eine Abhängigkeit der photokatalytischen Aktivität von der Partikelgröße nachweisen. Herr Striegler hat mit einem methodisch sehr breit angelegten Ansatz zahlreiche neue und hoch aktuelle Ergebnisse zur Modifizierung von Kohlenstoffnitriden für die photokatalytische Wasserspaltung erarbeitet. Die daraus gewonnenen Erkenntnisse stellen wichtige Impulse für zukünftige Arbeiten auf diesem Gebiet dar. Die schriftliche Masterarbeit ist nicht nur klar strukturiert und ansprechend gestaltet, sondern zudem in flüssig geschriebenem und klar verständlichem Englisch abgefasst. Die Argumentation ist trotz der komplexen Thematik verständlich dargestellt und die Kernergebnisse mit überzeugender Klarheit zusammengefasst.

Es freut mich sehr, dass die Masterarbeit von Herrn M.Sc. Karl Striegler, einem sehr engagierten und hoch motivierten jungen Wissenschaftler mit hohem Potential für eine weitere akademische Entwicklung, im Rahmen von Springer BestMasters 2014 prämiert wurde.

Prof. Dr. Roger Gläser

Institutsprofil

Das **Institut für Technische Chemie an der Universität Leipzig** verknüpft in Forschung und Lehre die Grundlagen der Materialwissenschaft mit technisch-industriellen Anwendungen. Im Zentrum der Anwendungen steht die Heterogene Katalyse. Die wesentlichen Anwendungsfelder sind Energie- und Umweltforschung. Die Präparation und Charakterisierung nanostrukturierter Materialien als Katalysatoren sowie die Nutzung nachwachsender Rohstoffe als Energiequelle oder für die Synthese von chemischen Zwischenprodukten stellen dabei Brückenthemen dar. Der Ansatz für ein vertieftes Verständnis der katalytischen Wirkungsweise basiert dabei auf der gezielten Herstellung und umfassenden physikalisch-chemischen Charakterisierung von Materialien mit definierten Eigenschaften einerseits und der Korrelation zu den katalytischen Eigenschaften unter anwendungsnahen Bedingungen andererseits. Neben funktionalen Materialien mit definierten Eigenschaften, v.a. hinsichtlich der Porosität auf unterschiedlichen Längenskalen, spielt die chemische Reaktionstechnik eine wesentliche Rolle als Werkzeug zur Interpretation der katalytischen Prozesse auf der Ebene des aktiven Zentrums in seiner Wirkungsumgebung. Damit dient die Reaktionstechnik der anwendungsorientierten experimentellen Untersuchung der Katalysatoren. Dieses Konzept verbindet somit materialorientierte Arbeiten mit der angewandt-ingenieurwissenschaftlichen Katalyseforschung einerseits sowie der homogen, heterogen und enzymatischen Katalyse andererseits.

In der Lehre werden Studierenden die Grundlagen der industriellen Chemie vermittelt und in Vertiefungsmodulen sowie Exkursionen die für technische Anwendungen und Prozesse wesentlichen Zusammenhänge und Perspektiven erarbeitet. Auf diese Weise werden die im naturwissenschaftlichen Studium erworbenen Kenntnisse mit den für die berufliche Tätigkeit im technisch-angewandten Kontext erforderlichen Fähigkeiten verbunden.

Die am Institut verfolgten wissenschaftlichen Zielsetzungen leiten sich unmittelbar aus dem Umfeld einer innovationsgetriebenen, nachhaltigen chemischen Industrie ab. Auf diese Weise wird ein Ansatz der experimentellen Heterogenen Katalyse und der angewandten Materialforschung auf der Grundlage der Technischen Chemie konsequent verfolgt.

Schlüssel für das Verständnis heterogen katalysierter Prozesse sind Katalysatoren mit definierten Eigenschaften. Neben der Nanostrukturierung von Katalysatoren v.a. hinsichtlich der Porosität wird ein breites Spektrum an katalytischen Aktivkomponenten untersucht. Die katalytisch aktiven Zentren sind dabei entweder als isolierte Atome in das Feststoffgerüst eines mikro-, meso- oder makroporösen Materials eingebaut oder liegen als Gastkomponenten, z.B. als Metallnanopartikel, in den Poren vor. In jüngerer Zeit wurden zudem Materialien untersucht, die als gesamte Phase aus einer katalytisch aktiven Komponente, z.B. aus Übergangsmetall(misch)oxiden, bestehen. Die Porendurchmesser der Materialien überstreichen einen weiten Größenbereich und gestatten so die Untersuchung skalenübergreifender Phänomene der Materialwissenschaft und der heterogenen Katalyse. Für solche Studien werden am Institut entwickelte Synthese von Materialien mit hierarchisch strukturierten Porensystemen durch kombinierte Templatierungsstrategien (Exo- und Endotemplate) oder Sol-Gel-Techniken konsequent genutzt und weiter ausgebaut. Mittels vielfältiger Methoden der physikalisch-chemischen Feststoffcharakterisierung wird der Zugang zum Verständnis von Struktur und Wirkungsweise nanoporöser Katalysatoren erreicht.

Zu den aktuellen Anwendungsgebieten der Materialien zählen neben adsorptiver Stofftrennung und Sensorik insbesondere die Umwandlung und Speicherung regenerativer Energie, z.B. die selektive katalytische Reduktion von Stickoxiden in Dieselabgasen im Betrieb mit Biokraftstoffen oder photo- und elektrokatalytische Routen zur Wasserstoffherstellung. Ein weiterer Schwerpunkt stellt die Nutzung innovativer Reaktionsmedien wie überkritischer Fluide und gasexpandierter Flüssigkeiten für die heterogene Katalyse dar.

Prof. Dr. Roger Gläser

Danksagung

*Für den Pudel und die Problematik mit dem Kern, beschrieben von Johann, irgend-
wann, auch hier, vermutlich auch für uns.*

*For the poodle and the problem with the essence,delineated by Johann, at some
point, here as well,probably also for us.*

Ich danke Professor Dr. Roger Gläser für die Möglichkeit meine Masterarbeit im Ar-
beitskreis „Heterogene Katalyse" und zur Photokatalyse machen zu dürfen. Das The-
ma hat mich nach den Monaten weitestgehend eingefangen und ein klein wenig
Überzeugungsarbeit geleistet, die wissenschaftliche Tätigkeit wieder lieb zu gewin-
nen. Die offene, freundliche Aufnahme, schnelle Unterstützung und eine gute wissenschaftliche Umgebung ist für dafür Bedingung. Vielen Dank dafür.
Natürlich würde ich nicht dazu kommen, eine Abschlussarbeit zu schreiben, wenn ich
nicht permanent dazu genötigt worden wäre, in die Schule zu gehen, fleißig zu studie-
ren und bei Unlust nicht zu verzagen. Für diese und viele andere Arten der jahrlangen
Unterstützung bedanke ich mich bei meinen Eltern. Eigentlich reicht es nicht, nur danke zu sagen oder zu schreiben, aber für mehr kann ich Papier leider nicht nutzen.
Dennoch danke.
Dennis Richter habe ich nicht nur die Beratung im Labor und für die Diskussion über
alle fachlichen und technischen Fragen, die mir immer sehr akut schienen, zu danken,
sondern auch für die allgemeinen Lebenslektionen. Außerdem hat er weitreichend
Analytik von photokatalytischen Experimenten bis TEM gewährleistet.
Mein Dank gebührt Professor Dr. Dirk Enke für die Übernahme als Zweitgutachter, die
er völlig natürlich, ohne große Bedenken übernahm und für die gute Laune, die er
stets vermittelt.
Bei chemischer Laborarbeit fallen immer wieder Analysen und Experimente an, die
man nicht alleine machen kann, deshalb sei in randomisierter Reihenfolge hier allen
gedankt, die mir einen Teil meiner Arbeit abnahmen und mich unterstützen: Sebastian
Zimmermann und Alexander Gnauck für die TPR-Messungen, Professor Dr. Wolf-
Dietrich Einicke für Stickstoffsorptionsmessungen, Juliane Titus für das Spiel mit dem
Feuer und den Mut g-C_3N_4 mehreren TG/DTA zu unterziehen, Gerd Kommichau für
schnelle XRD-Messungen, Kerstin Thiele und Heike Rudzik für die ICP-OES-
Messungen und die stete Diskussion, Dr. Agnes Schulze für die tolle Möglichkeit, das

Wissen und die Fähigkeiten des IOMs zu nutzen und die tolle Atmosphäre dort, Gunther Speichert für das Nanotracking, Andreas Ott für die große Menge an TEM-Aufnahmen, Dr. Jürgen Schiller und Dr. Claudia Birkemeyer für die Möglichkeit der Maldi-MS-Messungen und der konstruktiven Gespräche über die Spektren, Robert Eschrich und Thomas Rammelt für die Hilfe mit Werkzeug und Laborequipment, Dr. Stefan Zahn für meine Betreuung bei den DFT-Rechnungen, die ihn viel Zeit kosteten, die er geduldig und ausführlich ausführte.

Danke an das Labor 515 mit allen Studierenden und Promovierenden, die mich während meiner Masterarbeit begleitet und unterstützt haben. Generell hat sich die Arbeitsgruppe durch tolle Mitarbeiter_innen und Mitarbeit ausgezeichnet. Es ist schön, wenn man nach Hilfe fragt und sie einem auch gewährt wird. Es sind nicht alle erwähnt, aber es sei allen sehr herzlich gedankt.

Menschen Leipzigs, dieser Fakultät und all mein_e Freund_innen sind ein steter Quell der Freude und Partner_innen in zahlreichen, schweren und lustigen Diskussion jeglicher Art. Die Zeit hier wäre einiges grauer ohne euch großartige Menschen. Mirjam und Vanessa, ohne eure Unterstützung wäre alles zehnmal schwerer. Danke für die Zeit. Ihr seid fantastisch.

Karl Striegler

Abstract

Graphitic carbon nitrides, short g-C$_3$N$_4$, are a promising material for photocatalysis and catalysis in general. Since the discovery of their photocatalytic activity towards hydrogen evolution from water, recently, there have been many attempts to improve this material. This thesis was focused on material modifications and their characterization. A series of photocatalytic experiments was carried out. Different types of modifications and improvement of the basic material are presented.

Copolymerization of comonomers with the standard precursor was performed leading to a successful introduction of carbon and sulfur. The hydrogen evolution rate of the resulting material increased up to 150 % compared to the basic material. DFT calculations and DTA/TG were performed to examine this improvement.

Dye-sensitizing, a technique known to improve the performance of solar cells, was used to alter the absorption spectrum of g-C$_3$N$_4$. Covalent coupling of dyes resulted in a doubling of the photocatalytic activity while using just visible light.

Furthermore, new methods of nanoparticle deposition were tested for this material and the influence of nanoparticle size was examined. The used compounds platinum, rhodium and cobalt oxide were previously reported to promote the photocatalytic hydrogen evolution from water. TEM proved sufficient deposition of platinum and rhodium by well-controllable polyvinylpyrrolidone mediated synthesis on g-C$_3$N$_4$ resulting in a highly active photocatalyst. It was discovered that the hydrogen evolution rate depends on the size of Co$_3$O$_4$ nanoparticles.

The thesis was carried out between 2013-06-05 and 2013-11-12 at the Institute of Chemical Technology, research group Heterogeneous Catalysis of Prof. Dr. Roger Gläser.

Table of Contents

List of Abbreviations

2MP	2-Mercaptopyrimidin (4,6-Diamino-2-mercaptopyrimidine)
6MI	6-Mercaptoisocytosine(4-Amino-6-hydroxy-2-mercaptopyrimidine)
A/A⁻	Any redox couple
BA	Barbituric acid
BET	Brunauer-Emmet-Teller
CA	Cyanuric acid
DCA	Dicyandiamide
DIC	N,N'-Diisopropylcarbodiimide
DFT	Density Functional Theory
DMAP	4-Dimethylaminopyridine
DTA	Differential Thermo Analysis
ICP-OES	Inductive Coupled Plasma Optical Emission Spectroscopy
HOMO	Highest Occupied Molecular Orbital
HRTEM	High Resolution Transmission Electron Microscope
GC	Gas Chromatography
g-C$_3$N$_4$	Graphitic Carbon Nitride
g-C$_3$N$_4$ (MA)	Graphitic Carbon Nitride derived from Melamine
g-C$_3$N$_4$ (DCA	Graphitic Carbon Nitride derived from Dicyandiamide
g-CN$_{XX\_Y}$ (Z)	Colpolymerized Carbon Nitride from Z and X with a ratio of Y
l-Co$_3$O$_4$	large cobalt oxide nanoparticles
LUMO	Lowest Unoccupied Molecular Orbital
m-Co$_3$O$_4$	medium cobalt oxide nanoparticles
[M+H]$^+$	Molecular Mass plus 1 (plus one proton)
MA	Melamine
Maldi-MS	Matrix Assisted Laser Desorption Ionization Mass Spectrometry
NM	Noble Metal
NMR	Nuclear Magnetic Resonance
NP	Nanoparticle
Ppy	Polypyrrole
qE(A/A⁻)	redox potential of A/A⁻
R$_{stable}$	stabilizer for nanoparticles
RT	Room Temperature
s-Co$_3$O$_4$	small cobalt oxide nanoparticles
TEM	Transmission Electron Spectroscopy
TEOA	Triethylolamine
TCD	Thermal Conductivity Detector
TG	Thermogravimetrical (analysis)

TPR	Temperature Programmed Reduction
TTC	Trithiocyanuric acid
UV/Vis	Ultraviolet/visible
XPS	X-ray photoelectron spectroscopy
XRD	X-Ray Diffraction
WI	Wetness Impregnation

List of Symbols

Symbol	Unit	Meaning
2θ	°	Reflection angle
A_{BET}	$m^2\,g^{-1}$	BET surface
c	$mol\,l^{-1}$	Concentration
d	m	Distance
D	Gy	Dose
D_P	m	Particle diameter
D_V	$m^3\,g^{-1}$	Pore volume
$E°$	eV	Electrochemical potential
G	$J\,mol^{-1}$	Gibbs energy
H_e	-	Hartee energy
I	A	Electrical current
k	$mol\,h^{-1}$	reaction rate coefficient
m	g	Mass
MM	Da	Molecular mass
MW	$g\,mol^{-1}$	Molecular weight
m/z	$5.86 \cdot 10^{-9}\,g/C$	mass-to-charge ratio
n	-	Number of electrons
R	ohm	Ohmic resistance
r	$mol\,h^{-1}$	Reaction rate
T	$°K, °C$	Temperature
t	h, min, s	Time
U	V	Voltage
P	W	Power
V	l	Volume
β	°	half-width of a diffraction reflex
δ	nm	Overall reflected light
λ	m	Wavelength
η	V	Overpotential
χ	$mol\text{-}\%$	Mole fraction
ω	$wt.\text{-}\%$	Massfraction

1   Introduction and Objective

Hydrogen is an important basic chemical. It is widely used in industry for example methanol synthesis, the Haber-Bosch process, hydrogenation reactions or the Fischer-Tropsch process. Furthermore, over the last decades it has become evident that CO_2 is a greenhouse gas, which is presumably responsible for global warming. Thus and due to its limited availability, humankind will have to change from fossil fuels as primary energy sources to regenerative energy sources. H_2 is promising candidate as an energy carrier. Lately, research is focusing on pollution free hydrogen generation by photocatalytic water splitting or photoelectrolysis of water. These methods provide sustainable energy.

In the late 1970s *Fujishima* and *Honda* discovered that a photoelectrical cell in which the semiconductor titania acts as a photoelectrode, is able to split water while it is illuminated. Although there were huge efforts to establish a working system of different semiconductor types and developing new kinds of materials, there was no breakthrough in that field. Almost all semiconductors are lacking in stability against photocorroision; their overpotenial for hydrogen and oxygen evolution is too high; often, the band gap is not suitable and, very essential, they exhibit an inadequate efficiency for the photoelectrolysis of water.

Still, nanostructuring is a prerequisite for hydrogen evolution from water with suspended photocatalysts. Besides progress in sensitizing semiconductors and apply nanomaterials for suspended photocatalysts, recently, graphitic carbon nitrides or g-C_3N_4 were rediscovered, after *Liebig* synthesized a material he called "melon", a promising material nowadays. It showed a significant hydrogen evolution from water when it is irradiated. Graphitic carbon nitride can be simply synthesized by heating precursors like melamine or urea under nitrogen atmosphere above 500 °C. It was recently investigated with respect to its physical, chemical and catalytic properties. Still, the structure is not fully elucidated. The material has a special structure that makes it unique compared to other solid semiconductors. The smallest unit – a heptazine ring – probably forms a two dimensional network of carbon and nitrogen which can be compared to graphite.

Considering the presumed structure and the straight forward synthesis, a wide range of modifications are imaginable. Many attempts were carried out doping the semiconductor properties for photocatalytic hydrogen evolution from water by altering the structure and composition. A vast of methods was tested like vapour deposition techniques which are elaborate. But also copolymerization of barbituric acid was examined to alter the material. The advantages of g-C_3N_4 are defects of the structure which are free primary amine groups. Techniques from organic chemistry were applied to modify this material. Nevertheless, because of its chemical nature, graphitic carbon nitrides have supposedly a high overpotential for hydrogen and oxygen. It was reported that the reaction can be accelerated by the deposition of promoters for photocatalytic hydrogen evolution from water. It was the aim of this work to investigate new strategies to modify catalysts for photocatalytic hydrogen evolution from water. Therefore graphitic carbon nitride had to be altered by in-situ synthesis and post-functionalizing. All modifications were performed to alter the composition of the basic material and improve the semiconductor properties for hydrogen evolution or to reduce the overpotential. Overall, the basic catalyst was modified by copolymerizing the precursor with different compounds for a quick and easy shift of the valence band. Here, several copolymers were used to introduce heteroatoms like sulfur and oxygen to investigate the influence of additional carbon atoms. Furthermore the polymeric carbon nitrides were sensitized with a dye for a wider absorbance of the visible light spectrum. Two basic techniques for immobilization were tried. It was tested whether it is possible to deposit the dye on the surface by electrostatic adsorption or to bind a dye covalently, respectively. In order to lower the overpotential for H_2 and O_2 evolution new ways of depositing nano-sized promoters like platinum and rhodium were investigated. Furthermore, a size dependency of the promoters Co_3O_4 was examined. The products were analyzed, as to whether syntheses had been conducted successfully or not.

To study the copolymerization process and the structure of graphitic carbon nitride XRD, classical elemental analysis, TG and DTA, Maldi mass spectrometry and DFT calculations of several HOMOs and LUMOs were perfomed. Successful dye sensitizing was investigated by diffuse reflection UV/Vis spectra. However, additionally experiments for the deposition of nanoparticles used nanoparticle tracking analysis, TEM and TPR. Of course, all synthesized materials were tested for their performance in photocatalytic hydrogen evolution from water, to deduce a relationship between a specific modification and its efficiency.

2 Literature Overview

2.1 Water Splitting

2.1.1 Hydrogen from Convenient Water Splitting

Hydrogen is required for several applications. Besides classic chemical use like the Haber-Bosch process, coal liquefaction or for hydrogenation, it is to be considered as an energy carrier.[1,2] Ever since hydrogen has been used on a large scale, there has been the need for an efficient, cheap hydrogen production.[3] In the past, numerous ways of hydrogen production were developed. Still, fossil fuels are regarded to be the prerequisite for steam reforming.[4] Furthermore, partial oxidation or coal gasification are also depending on fossil fuels.[5] Recently, biomass reforming has become more and more attractive. Obviously, hydrogen and oxygen can be produced by water splitting. While several methods are possible (e.g. thermal decomposition of water or photobiological water splitting), electrolysis is the most common. With around 5 % of the overall hydrogen production it is still a minor issue for industrial application compared to steam reforming.[6] Nevertheless, photocatalytical water splitting gained attention, because it converts the primary energy light into a chemical energy carrier, e.g. hydrogen.[7] It is proposed that hydrogen could alternatively be used as an energy-rich reagent for formation of hydrocarbons using atmospheric CO_2.[2]

$$H_2O \longrightarrow H_2 + 0.5\,O_2 \qquad (2\text{-}1)$$

$$2\,H_2O + 2\,e^- \longrightarrow H_2 + 2\,OH^- \qquad (2\text{-}2)$$

$$2\,OH^- \longrightarrow 0.5\,O_2 + H_2O + 2\,e^- \qquad (2\text{-}3)$$

Given that equation (2-1) is clearly an endergonic reaction with ΔG = 237 kj mol^{-1} = 1.23 eV. It is a requirement that some sort of energy must be fed to the system. In order to understand the ongoing research in photocatalytical and photoelectrochemical water splitting, the fundamentals of water electrolysis must be understood. While reaction (2-2) is the cathodic reduction and (2-3) is the anodic oxidation of water in alkaline electrolysis, it should be noted that the voltage ΔU to drive an open electrochemical cell is given in equation (2-4). Anyhow, two electrons have to move from the cathode to the anode.

$$\Delta U = \frac{\Delta G}{nF} + IR + \Sigma\eta \qquad (2\text{-}4)$$

While the number of electrons n, the Faraday constant F and the free Gibbs energy ΔG are set by the reaction itself, the current I, the total ohmic series resistance R and the overpotentials η can be influenced by designing the electrodes or a possible photoelectrical catalyst. [3] Hence, it has to be considered that oxygen can have an over-potential of about one volt for elemental metals[8], while the hydrogen overpotential is about half a volt for graphite.[9] The electrodes must not only be conducting but also have a low overpotential for hydrogen and oxygen. Of course, an increased current is favorable due to a higher reaction rate and can be achieved by a higher contact be-tween electrodes and the liquid. Nevertheless, the ohmic series resistance is sup-posed to be kept low.

Because the kinetics of hydrogen evolution from water electrolysis and the kinetics of photocatalytic hydrogen evolution from water can supposedly be compared, the mechanism and the kinetic will be discussed in the following.

$$H^+_{aq} + e^- \longrightarrow H_{ads} \qquad \text{Volmer} \qquad (2\text{-}5)$$
$$H_{ads} + H^+_{aq} + e^- \longrightarrow H_2 \qquad \text{Heyrowsky} \qquad (2\text{-}6)$$
$$2\,H_{ads} \longrightarrow H_2 \qquad \text{Tafel} \qquad (2\text{-}7)$$

The equations (2-5) to (2-7) describe the possible mechanisms of electrochemical hydrogen evolution. At first the hydrogen has to be adsorbed and reduced on the solid surface (2-5, Volmer mechanism). The second step depends on the affinity of the electrode towards hydrogen adsorption. For a low affinity equation (2-6) – Heyrowsky mechanism – is favored. For a high affinity equation 2-7 – Tafel mechanism – is pre-ferred. Therefore there are two possibilities of the overall proton reduction: Volmer-Tafel and Volmer-Heyrowski. For metals, the Volmer mechanism is most commonly the rate determining step. It is not surprising that metals with a medium metal-hydrogen bond energy inhibit neither Hyrowsky-mechanism nor the Tafel-mechanism.[10] Platinum, palladium and other noble metals are known to have low elec-trode overpotential. Thus, they have a medium metal-hydrogen bond energy, which leads to a fast Volmer-mechanism reaction, resulting in a quick reaction at rate deter-mining step.

Electrolysis of water or photocatalytic water splitting is always a zero-order reaction. A zero-rate reaction is found, when the reaction rate is independent on the concentra-tion of the reactants. This is the case for reactions, when the reaction rate depends on catalysts saturated with reactants. The reaction rate r_A is constant over time t, be-cause the increase or decrease of the concentration dc_A is constant; giving a system that depends only in the reaction rate coefficient k_A (equation 2-8). Since the diffusion to the surface is faster than the rate determining Volmer reaction, the concentration of the adsorbed H_{ads} will remain constant over the reaction time. This is the case for wa-ter splitting depending only on the surface of the catalyst and the applied bias. By al-tering the latter variable, k_A can be changed, but not the reaction order. Of course, photocatalytic reactions are special, because they occur mostly only under irradiation.

While changing from electrolysis of water to photocatalytic water splitting, the origin of the bias changes instead of the reaction order.

$$r_A = \frac{dc_A}{t} = k_A \qquad (2\text{-}8)$$

2.1.2 Photocatalytic Water Splitting and Hydrogen Evolution from Water

In general, photocatalysis can be divided in three groups. The excitement or stabilization of a reactant leads to a catalytic reaction what is characteristic for catalytic photoreactions. The reaction would not occur without light or catalyst. In photosensitized and photoassisted catalytic reactions the catalyst itself is required to be excited. For photosensitizing just an initial irradiation is required, after which the reaction will proceed even in the dark. Photoswitches are a typical example for this reaction type. Photoassisted reactions require a permanent irradiation of the catalyst, because the received light energy of the catalyst is transferred to the reactant during the reaction. The provided energy is a prerequisite for the reaction of the reactant or reactants. Therefore, photocatalytic water splitting or hydrogen evolution from water can be described as photoassisted catalysis.[11]

While various methods for water splitting from light are imaginable and applicable, photoelectrosynthetic cells (PECs) and suspended photocatalysts are the most encouraging challenges.[12] The energy conversion efficiency from sunlight to hydrogen is low. Still, 16 % efficiency of a commercial available photovoltaic cell plus electrolysis would exceed 12 % efficiency of the best PECs in laboratory scale.[13] Photocatalytic water splitting or photocatalytic electrolysis of water was discovered by *Fujishima* and *Honda*.[14,15] They established illuminated TiO_2 as the material for "artificial photosynthesis". Though, in these experiments TiO_2 was used as the photoanode, while platinum was used as the metal cathode, it was postulated, that colloid systems would also split water.[16,17] Nevertheless, a first commercial attempt of using titania for solar energy conversion was done by *Grätzel*: a combination of nanocrystalline TiO_2 and certain dyes supported on a transparent conducting oxide as anode.[18]

As described in section 2.1.1, water can be split by electrical current. Photocatalytic water splitting is achieved by a photogenerated electron and an electron hole within a semiconductor. While the potential of an excited electron, which was promoted to the conduction band, has to be lower than 0.00 V, the remaining hole in the valence band must have a potential above 1.23 V in order to split water into its elements. The band edge energies must straddle the electrochemical potentials of $E°(H^+/H_2)$ and $E°(O_2/H_2O)$ to drive the hydrogen or oxygen evolution reaction, respectively. This is illustrated in Figure 2-1a.[19] Thus, in terms of thermodynamics photocatalytic water splitting is not very extraordinary. Since the free enthalpy of the reaction is 1.23 eV, a semiconductor material is needed whose absorption edge is above that energy.

Hence, the wavelength of light used for water splitting is required to be less than 1010 nm. Though this would cover the whole visible and a part of the near infrared spectrum, the overpotential for the oxidation and reduction of water inhibits photoelectrolysis for the most semiconductors. This is observed, even if an appropriate catalyst is used for the generation of electron-hole pairs. Therefore, the required band gap of a semiconductor for water splitting is reported to be between 1.6 eV and 2.4 eV.[20] Anyhow, working photochemical electrodes can be designed in two configurations: Schottky type cells have a photoanode-cathode configuration and Tandem or Z-Scheme type cells use a photoanode-photocathode configuration. The latter have the advantage of two smaller band gaps instead of a single, large one. Thus, a greater fraction of the solar spectrum can be used.[12] Though, when it comes down to use suspended photocatalysts, just Schottky type cells seem to have reasonable efficiency.[21,22] This might be because of the difficult synthesis of two microscale semiconductors particles, which have an ohmic contact.

After the demonstration of *Fujishima* and *Honda* of artificial photosynthesis, photocatalytic effects were demonstrated on suspended semiconductor particles.[14] Furthermore, it was assumed that the basic functions of a "photoelectrosynthetic" cell could be imitated by nano- or microscale objects. Light absorption, charge separation and water electrolysis is supposed to be accomplished by one particle instead of two electrodes. The principle is illustrated in Figure 2-1b.[23] There were early attempts with colloid TiO_2 structures which showed no overall water splitting. As it is depicted, after the generation of an electron-hole pair, the migration to the surface is a prerequisite for the catalytic reaction. Crystal structure, crystallinity and particle size can clearly influence migration. Crystal defects may affect the lifetime of an electron-hole pair, since they can recombine at these centers. In smaller particles the way to the reaction sites becomes shorter, which leads to a decreased probability for electron-hole recombination.

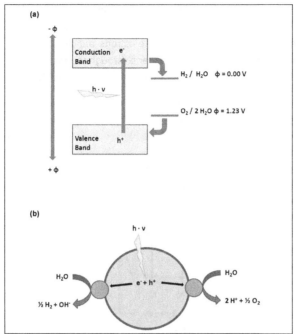

Figure 2-1: *Principles of photoelelectrolysis of water. (a) Schematic illustration of the band gap prerequisite of semiconductor for water splitting. (b) Model for the charge separation by irradiation within the photocatalyst and hydrogen plus oxygen evolution by deposited promoters.*

Furthermore, the reactive surface character and the number of active sites are important. It was repeatedly reported, that sufficient photocatalytic water splitting was observed only in the presence of deposited nanocomposites for some semiconductors catalyzing the oxygen evolution reaction, the hydrogen evolution reaction or both. Although for the most metal oxide a promoter is not needed for oxygen production, still, hydrogen evolution is depending on typical promoter for hydrogen reactions like Pt, Pd or Rh.[7,19] Considerations of the nanostructuring of catalyst and promoters can be found in section 2.1.3.

Although a suitable band and its position are the requirement for photoelectrolysis of water, the energetics of semiconductor liquid interface under irradiation are crucial for the understanding of that process. The actual free energy, which is basically the Fermi energy of the semiconductor, is influenced by the kinetics of the charge carriers under irradiation. Therefore the effective band gap under illumination is smaller than in the dark. Hence, water cannot be split by a catalyst under irradiation unless the photovoltage is less than 1.23 V.[24] Kinetically, the formation rate of charge carriers must be higher than their consumption rate. Therefore, five non-positively contributing effi-

ciency processes can be outlined. These are illustrated in Figure 2-2. The photogenerated electron-hole pair is able to recombine in the bulk (1), in the depletion region
(2) or at surface defects (3). Furthermore, electrons are able to tunnel through the
electric barrier near the surface (4) and escape through thermionic emission.[12] Thus,
it should be the aim developing well organized semiconductors for inhibiting the processes (2) to (5).[25]

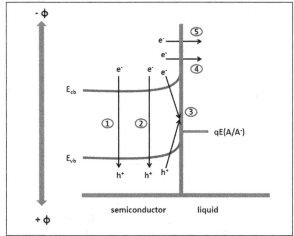

Figure 2-2:   *Schematic illustration of the recombination pathways of photogenerated electron-hole
pairs within a semiconductor under irradiation. The semiconductor is in contact with a liquid and
qE(A/A⁻) marks the electrochemical potential for the redox couple of acceptor A and donor A⁻.*

Semiconductors with insufficient water splitting abilities might be able to split water
with a chemical bias such as sacrificial agents being electron donors or acceptors.
Hence, hydrogen or oxygen evolution rates are enhanced depending on the nature of
the sacrificial agent. On the one hand oxidizing agents like Ag^+ or Fe^{3+} were able to
increase the oxygen evolution rate while hydrogen production was suppressed; on the
other hand alcohols or thiols achieved the opposite. Sacrificial agents or chemical bi-
ases, as they are also referred, are crucial nowadays for sufficient for photocatalytic
hydrogen evolution from water. This is a huge drawback of this method. Without these
agents the hydrogen evolution rate of photocatalyzed water splitting is always close to
zero. There were very rare examples of overall water splitting into hydrogen and oxygen. Still, it will be one of the major challenges in the next decades to overcome this
problem. Most commonly methanol or triethanolamine is used to get oxidized instead
of the oxygen within the water. The lower redox potential of the organic sacrificial
agents was discussed to be the major influence for the enhancement of the photo-

catalytic activity. Also the high overpotentials for oxygen oxidation might have a large effect on a suppressed hydrogen evolution.[26]

$$\text{CdS} \longrightarrow \text{Cd}^{2+} + \text{S} + 2\,\text{e}^- \qquad (2\text{-}9)$$

Besides all development in photocatalysis, photocorrosion is still the biggest disadvantage of using light as a primary energy source. For example, CdS has been studied for a while and it has a suitable band gab and band gap position, but instead of oxidizing water, the catalyst itself is oxidized without a matching sacrificial agent.[27] The reaction shown in equation (2-9) can replace O^{2-} oxidation shown in equation (2-3). Although the degree of photocorrosion depending on the used material, it will remain certainly as the main drawback in all kinds of photocatalysis, homogenously or heterogeneously.

Because nowadays it is state of the art to develop suspended photocatalysts for hydrogen evolution from water, the discussion have to outline the advantages of nanostructured catalysts. Since it became clear that nanostructuring is a prerequisite for the conceptual transition from photoelectrochemical cells to suspended photocatalyst, this quantum effect was used and examined.[28] This will delineated in section 2.1.3. Nevertheless, even early researches focused on titania nanoparticle photoanodes.

2.1.3 Nanostructuring in Photocatalysis for Sufficient Hydrogen Evolution

Since the groundbreaking talk of physicist Richard Feynman "There is Plenty of Room at the Bottom" in 1956,[29] nano-technology has had a huge impact on industrial, biomedical, electronic and catalytic applications. This is to take advantage of their small size, resulting in higher specific surface area and properties that are not observed in bulk material like changes of the band gap or ion conductivity. Nanoparticles (NPs) are defined by IUPAC to have dimensions between 1 and 100 nm. NPs can be synthesized by "top down" strategies using vapor deposition etc. or "bottom up" approaches, utilizing classically chemical strategies.[30,31]

$$x\,\text{M}^{n+} + \text{R}_{stable} + nx\,\text{e}^- \longrightarrow x\,\text{M}^0_{n\ stabilezed,\ cluster} \qquad (2\text{-}10)$$

Soft template design became most common for metallic nanoparticles. Therefore, "wet chemical" reduction has proven to be a good method for the preparation of solvent dispersed NPs.[32] As shown in equation (2-10), the general procedure is straight forward. A metal salt M^{n+} is dissolved with a stabilizer R_{stable} and a reducing agent herein implied as a solved electron e^-. For the reduction of metal salts, several compounds were found to be suitable like hydrogen, borohydrides, alcohols or hydrazines. While the metal is reduced, the stabilizer inhibits growth of the metal nucleus above a certain size and promotes a colloid solubility of the metal cluster.[30] Additionally, the size of the NPs can be controlled.[33]

However, the usage of metal NPs for catalysis has been investigated thoroughly. Since the reaction rate for most typical metal catalyzed reaction increases due to a larger specific surface area, there were also investigations about shape- and size-depending effects for these reactions when using catalytically active nanoparticles. There are cases for metallic systems in which the activity goes up or descends with smaller NPs while the surface area is held constant. These effects are called negative or positive size effects. Nevertheless, there are cases in which the activity goes through a maximum of a certain NP core diameter.[34]

Since there were concepts to apply nanoscaled particles on sufficient water splitting, a lot of research was conducted to improve the catalytic activity of suspended nanoparticles. Nowadays, it is proven that only nanostructured photocatalyst can provide a sufficient hydrogen evolution rate from water when a suspended catalyst is used instead of a photoelectrochemical cell.[35] The ability of suspended semiconductors to split water is based on the quantum size effect what was already denoted in Figure 2-1 b. Size and electronic properties play an important role. In a solid bulk catalyst is no electron-hole pair separation possible. The pure volume and defect sites of the semiconductor material do not allow the electrons or the positively charged holes to diffuse to the solid surface without recombination.

Though it seems to be obvious that catalytic activity for photocatalytic water splitting can be enhanced by smaller particles due to a higher specific surface area, pros and cons of nanostructures have to be discussed. The following discussion focuses mainly on kinetic aspects regarding light distribution and electron diffusion.[19]

For water splitting, photoexcited charge carriers are supposed to diffuse to the surface. Because their lifetime is in the range of microseconds, a nanostructured surface or a small NP increases the probability of these electronic states to reach the semiconductor surface. Similar considerations might be done for the light distribution. The degree of horizontal light distribution will be enlarged by light scattering. In general, this effect can clearly be seen by UV/Vis absorption measurements of colloidal suspensions. Certainly, quantum size confinement is most important regarding photocatalytic experiments using a semiconductor. *Holmes et al.* observed a logarithmic size-dependency in catalysis using CdSe quantum dots.[36] That can be explained by a growing band gap when the size of the particle decreases. Derived from the Marcus-Gerischer theory increasing thermodynamic driving force accelerates the reaction for electrolysis.[37] All these effects have been applied to photoelectrolysis. Besides all positive effects, there are negative contributions from NPs for photocatalysis. As implied in Figure 2-2, a higher specific surface area allows a higher surface electron recombination rate. Hence, the quantum efficiency decreases. Intuitively, in smaller particles the space for charge carrier separation is reduced. The charge layers are not effectively isolated and additional energy would be required.

However, two major disadvantages of nanostructuring are a lower photon flux at surfaces leading to a lower flux to specific surface area ratio and a slow interparticle charge transport due to carrier diffusion instead of carrier drift. All the pros and cons will have to be considered to outline a good strategy for heterogeneous photocatalysis.[19]

Although early results of $Grätzel$ neglected a satisfactory water splitting reaction, recently, microscaled suspended photocatalysts worked under ultraviolet conditions without a chemical bias. Nevertheless, all of them are Schottky type devices, but only a few compounds showed overall water splitting.

There were inter alia transition metal compounds like RuO_2 modified $LiNbO_3$.[38] Water splitting is complex, kinetically slower, more corrosive and highly endergonic compared to a single-electron-transfer. Fe_2O_3, WO_3 or MnO_2 have been developed as nanostructured photoelectrodes which might be used for a bias-free photoelectrolysis.[39] These materials were improved with small promoters like NiO, IrO_2 or MoS_2. The latest research focuses on several techniques and materials, in order to tailor the properties and the catalytic activity. As already mentioned, it is the overall aim to develop nanostructured photocatalyst in suspension. Most common are metal oxides in general. ZnO, TiO_2 and Fe_2O_3 are very promising, because they have a suitable band gap for water splitting. Although cobalt modifications of these materials were performed to enhance the photocurrent, they lack in catalytic activity towards water splitting experiment.[16,40] The performance is better with materials containing niobates, which would be expensive in a large scale. Furthermore, nickel and tantalum catalyst showed recently a reasonable activity.[41] Still, photocorrosion, the difficult synthesis and the costs of these systems will remain a major drawback of metal oxides. Unfortunately, the most common semiconductors with a suitable band gap are metal chalcogenides which underlie the same processes as the described photocatalysts. Recently, graphitic carbon nitride was discovered as a photocatalytic active semiconductor for hydrogen evolution, which has shown none of the disadvantages so far.

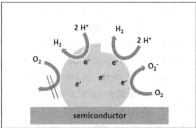

Figure 2-3: *Illustration of the effect of a Cr$_2$O$_3$ shell surrounding a noble metal cocatalyst on an illuminated semiconductor-photocatalyst.*

The benefit of promoters for electrode reactions have been known for 40 years.[42] More specifically, electron injection from TiO$_2$ into gold [NPs] and hole injection into RuO$_2$ was directly observed by transient absorption spectroscopy.[43] Furthermore, Pd, Ru, Rh and Pt are well studied for their properties as cocatalysts for water reduction reactions.[22] For photocatalytic water splitting, the reaction rate becomes equal to the rate of the back reaction at certain H$_2$ and O$_2$ concentrations. Therefore, *Domen et al.* elucidated that a 2 nm layer of Cr$_2$O$_3$ on the cocatalyst catalyzes a proton reduction but inhibits reduction of the generated oxygen. That is illustrated in Figure 2-3.[44] Nevertheless, there have been also reports about several oxides promoting the oxidation at photoactive semiconductors like IrO$_2$ or different cobalt oxides. This might be attributed to the reduction of the overpotential.[45] Investigations on promoters for photocatalytic hydrogen evolution from water will be crucial for the development and success of that method.

2.2 Graphitic Carbon Nitrides

2.2.1 Structure

The history of graphitic carbon nitrides reaches back to *Liebig* who discovered the so called "melon".[46] Until 1990, this class of compounds was not attractive for research.[47] It was the theoretical prediction of β-carbon nitride which led to experimental activities.[48] Nowadays, there are many efforts toward synthesis and characterization of graphitic carbon nitrides.[49] In general, graphitic carbon nitrides, a class of materials with a carbon nitrogen ratio close to three to four, are often referred as g-C$_3$N$_4$. Although there have been many attempts, this compound could not be crystallized well enough for X-ray crystal analysis.[50,51] It is said, that nanostructuring, template chemistry and copolymeric modification are supposed to be more important for material design than growing of single crystals. In 2006, this class of materials was considered to be used in metal-free catalysis.[52]

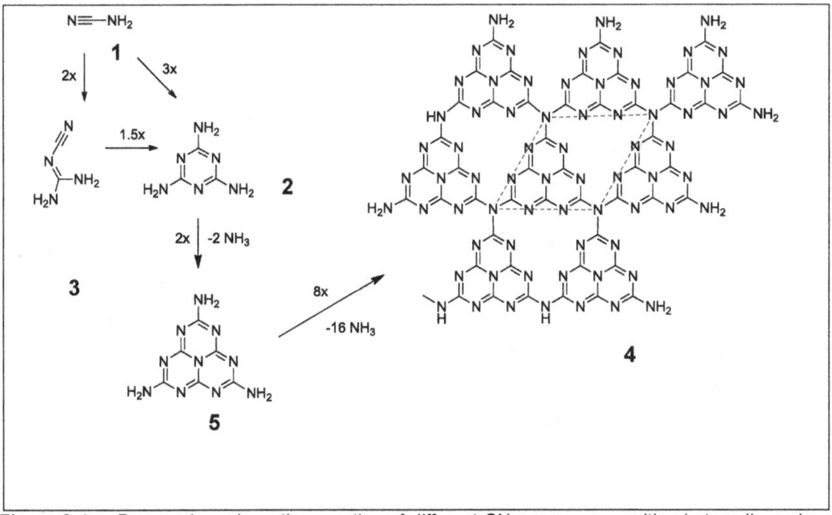

Figure 2-4: *Proposed condensation reaction of different CN$_x$ precursor resulting in two-dimensional sheets of graphitic carbon nitrides. The unit mesh of g-C$_3$N$_4$ is shown as as dashed parallelogram.*

Since the synthesis of graphitic carbon nitrides has been studied very well,[53] a lot of strategies have been developed. Starting from several materials, about 20 different preparation methods and precursors were developed sufficiently.[49] Main synthesis routes are depicted in Figure 2-4. One of the easiest procedures is heating cyanamide (**1**), dicyandiamide (**3**) or just melamine (**2**) under nitrogen atmosphere up to 500 °C - 600 °C. This starts a polymerization process that ends with graphitic carbon nitride (**4**).[54] However, beginning with the polymerization of cyanamid is not a prerequisite for a successful synthesis.[55] The synthesis of g-C$_3$N$_4$ can be started from dicyandiamide and melamine. Despite the fact, that starting materials are different, when it comes down to analyze the intermediates, it is known, that melamine is always the main intermediate, which all synthesis routes have in common. Nevertheless, melam (not shown) and melem (**5**) were also identified intermediary products.[56] Compound **5** is also referred as a heptazine or tri-s-triazine derivative. Though the carbon nitride is illustrated as a polymer of eight heptazines, it is proposed that hundreds of these units can condensate to form large sheets. In that case, a unit mesh, indicated with red dashes in Figure 2-4, would continue two-dimensional for an undefined length. It is worth noticing, that these so called sheets are a distinct π-system. A structure, which is defined but still has amorphous components, was discussed for a long time. There are contributions for structure elucidation by HRTEM and coupled X-ray powder spectra, implying that those graphitic carbon nitrides have a broken hexagonal symmetry as expected. This is, the structure is comparable with the structure of graphite consisting of stacked sheets.[57] Nevertheless, it was discovered early that the hydrogen mass fraction χ_H varies between 1.1 and 2.0 mol.-%[58] Therefore, the unit mesh must be interrupted quite often and free primary and secondary amine are available for modification.

There were many efforts to test the stability of graphitic carbon nitrides and investigate new modifications of that material. The precursors are organic and the compound itself seems to be an organic polymer having free amine functions. So, thermal decomposition, breaking up by solvents or modifications seem to be very likely. Surprisingly, thermogravimetrical analysis has proven a good thermal stability up to 630 °C. At this temperature the material decomposes without leaving a residue. Hence, it is more temperature stabile than so called high temperature polymers. Furthermore, it was shown that graphitic carbon nitrides are stable in all common solvents like water, ethanol, acetone and many others.[59] Using strong acids, it can be protonated permanently and form colloid solutions. Using an alkaline melt, it is hydroxylized.[60] Nevertheless, it is almost insoluble in water or organic solvents.

2.2.2 Optical and Photocatalytical Properties

Graphitic carbon nitride was very well characterized by means of UV/Vis spectroscopy and photoluminescence experiments. A regularly described yellow to orange color of this semiconductor matches with a band gap of 2.7 eV. Theoretical calculations showed that the absorption edge depends on structure packing and adatoms. Experimental data confirms an absorption edge at about 430 nm.[49] Different preparation methods are able to alter the optical properties. Mesoporous designed material was found to have a higher photocurrent than solid graphitic carbon nitride. The benefit of modifications will be described in detail in section 2.2.3.

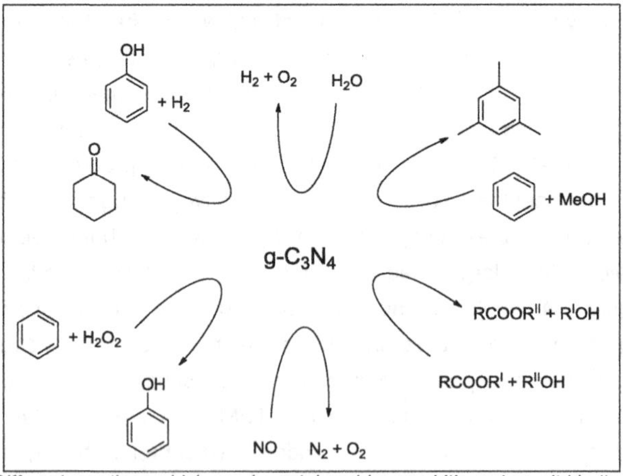

Figure 2-5: *Different reactions which can be catalyzed by graphitic carbon nitride itself or in which it has an important role as a support.*

Although g-C$_3$N$_4$ chemistry was just recently rediscovered a variety of catalytic applications has emerged (Figure 2-5). Friedel-Crafts acylation is a typical reaction of activated aromatic rings. It was shown that mesoporous graphitc carbon nitride was able to catalyze this reactiontype. Electron density might be transferred to the aromatic ring.[61] Furthermore, it could be used as a basic catalyst in transesterification[62] and as a non-noble metal catalyst with a low turnover in the decomposition of NO[63] Interestingly, it can be also used for alkene oxidation and alkene hydrogenation.[64] Being an easily synthesized semiconductor for photocatalytic water splitting is the most encouraging property of this material. This is remarkable since it is a metal free photocatalyst for water splitting. This might be a prerequisite for cheap, large-scale application for that kind of hydrogen production.

As expected, the semiconductor graphitic carbon nitride has a suitable band gap and band gap position. The energy of the valence band straddles the potential for water oxidation and the energy of the conduction band the potential for water reduction. Of course, the band gap of 2.7 eV exceeds the energy of 1.23 eV and the experimental often found kinetic overpotential of semiconductors of about 1 eV.[65] Furthermore, it has the fitting microstructure that can bind molecules by surface minimization, defects and nitrogen atoms. Nevertheless, still two major modifications have been regularly reported to be required for a sufficient photoelectrolysis of water. Although the semiconductor energy levels straddle redox potentials of water, still sacrificial agents are needed for water splitting. Triethanolamine and methanol were reported to be good additives.[47] Still, a noble metal promoter is needed for these reactions, due to the overpotentials of both partial reactions. As already discussed in section 2.1.2, hydrogen evolution rate was enhanced by deposited platinum, rhodium or palladium. For other semiconductors, deposition of noble metals led to an increased oxygen evolution rate. Cr$_2$O$_3$, Co$_3$O$_4$ and Co$_3$(PO$_4$)$_2$ were applied for that purpose.[66,67]

2.2.3   Doping Photocatalytic Properties

As already mentioned in section 2.2.1, g-C$_3$N$_4$ can be modified by several approaches. Besides structure variation, it can be either functionalized post-synthesis or in-situ using different synthesis strategies. Texture modifications were carried out in order to increase the specific surface area. Therefore nanocasting was performed with mesoporous silica matrices as templates. Hence, specific surface area and hydrogen evolution rate has been enlarged by one order of magnitude.[68] Chemical modification for various semiconductors have been studied and applied for a while. As implied by the main syntheses strategies, graphitic carbon nitride was doped with several elements. Typically, doping g-C$_3$N$_4$ involves copolymerization of precursor such as dicyandiamide or melamine with other comonomers. There have been successful efforts doping this material with boron and phosphorous using ionic liquids as doping agent.[69]

The resulted material showed improved catalytic activity but was not tested in photo-catalysis.

There have been three chemical modifications that showed improvements for water splitting without promoters: sulfur doping, peptizing with hydrochloric acid and copol-ymerization with barbituric acid.[70] Sulfur doping was achieved by treating well pol-ymerized g-C_3N_4 with hydrogen sulfide at 450 °C. The resulting, well characterized material showed variable sulfur content and a good water splitting performance. In sulfur-mediated synthesis the energy level of the HOMO is lowered, making the hole in the valence band a stronger oxidizing agent.[71] Furthermore, a synthesis of sulfur doped carbon nitride was done by polymerization of trithiocyanuric acid.[72] Because the precursor is highly volatile, the reaction need to be conducted under inert atmos-phere. Colloid graphitic carbon nitride was easily produced by heating it in hydrochlo-ric acid under reflux. Spectroscopic data implied that hydrochloric acid acts as an ada-tom, so the configuration might change and the acid intercalates between the stacked sheets. Nevertheless, this modification is reversible when heated above 350 °C.[73] Barbituric acid was effectively copolymerized with melamine. This was studied to change the carbon to nitride ratio from three to four to higher carbon content, disrupt aromatic system partially and introduce more crystal defects. The absorption edge of g-C_3N_4 was efficiently red-shifted. A higher absorption in the visible region of light promised a higher energy conversion and increased hydrogen evolution rate. This was verified with weakly doped basic material.

Since the discovery of titania as a photocatalytic material, Grätzel et al. tried to estab-lish a solar cell with TiO_2. They immobilized certain dyes on titania-nanoparticles which were coated on a transparent electrode, in order to broaden the range of light that can be used for the conversion of solar energy. These included metal complex dyes with ruthenium as well as metal-free dyes like erythrosine B. A very schematic illustration describes the process for photoelectrochemical cells in Figure 2-6. A solved redox couple transports electrons from the cathode to the dye-sensitized an-ode. The energy of an excited electron gets transferred to the semiconductor.[74] Alt-hough graphitic carbon nitride has not been used for solar cells, only in suspended photocatalytic experiments, it was confirmed that magnesium phthalocyanine could be deposited at the semiconductor. Hence, higher quantum efficiency was achieved for photoelectrolysis.[75]

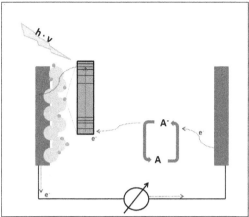

Figure 2-6: *Schematic illustration of an operating dye-sensitized solar cell. Pathway of the electrons (red) within the cell from the cathode over an electrolyte redox couple (A/A⁻) for electron transport and energy uptake by the dye which (violet) conducts it over titania (grey) to the transparent anode.*

3 Experimental Section

3.1 Equipment

Table 3-1: Applied equipment for measurements, including name of model and providing company

(Analytical) Method / Purpose	Model	Company
Electron Beam Modification	ELSA	Institut für Oberflächenmodifizierung
CNS Elemental Analysis	Vario IL	Elementaranalyse Systeme Hanau
Gas Chromatography	HP 6890 Series GC Systems	Hewlett Packard
ICP-OES	Optima 8000	Perkin Elmer
Light Source	Hg medium pressure lamp	Heraeus
Maldi-MS	Daltonics	Bruker
N_2-Sorption	Asap 2010	Micromeritics Instruments
Nanoparticle Tracking Analysis	LM10	NanoSight
Thermal Analysis	STA 409	Netzsch
TEM	Jem 2100	Jeol
TPR	Ami 100	Altamira Instruments
UV/Vis	Lambda 650 s	Perkin Elmer
XRD	XRD-7	Zeiss

3.2 Chemicals

Table 3-2: Overview of used chemicals, their formula, molecular weight and commercial producer

Compound	Formula	MW / g mol^{-1}	Company
4,6-Diamino-2-mercaptopyrimidine	$C_4H_6N_4S$	142.2	Sigma-Aldrich
4-Amino-6-hydroxy-2-mercaptopyrimidine	$C_4H_5N_3OS$	161.2	Sigma-Aldrich
4-Dimethylaminopyridine	$C_7H_{10}N_2$	122.7	Fluka
Acetonitril (for HPLC)	C_2H_3N	41.1	VWR
Ammonia (25 %)	NH_3	17.0	Merck
Barbituric acid	$C_4H_4N_2O_3$	128.1	Bernd Kraft
Cobalt (II) acetate tetrahydrate	$Co(OAc)_2 \cdot 4H_2O$	249.1	Merck
Cobalt (II) chloride hexahydrate	$CoCl_2 \cdot 6H_2O$	237.9	Merck
Cyanuric acid	$C_3H_3N_3O_3$	129.1	Sigma-Aldrich
Dicyandiamide	$C_2H_4N_4$	84.1	Sigma-Aldrich
Erythrosine B	$C_{20}H_6I_4Na_2O_5$	879.9	Abcr
Ethanol (absolute, 99.9 %)	C_2H_6O	46.1	VWR
Ethylacetate (99 %)	$C_4H_8O_2$	88.1	Sigma-Aldrich
Ethylene glycol (99 %)	$C_2H_6O_2$	62.1	Sigma-Aldrich
Hexane (chemically pure)	C_6H_{14}	86.2	Merck
Hexachloroplatinic acid hexahydrate	$H_2[PtCl_6] \cdot 6H_2O$	517.9	Private
Hydrochloric acid (37 %)	HCl	36.5	Merck
Melamine	$C_3H_6N_6$	126.1	Alfa Aesar
Methanol (chemically pure)	CH_4O	32.0	Merck
Nitric acid (65 %)	HNO_3	63.0	Merck
N,N'-Diisopropylcarbodiimide	$C_7H_{14}N_2$	126.2	Sigma-Aldrich
Polyvinylpyrrolidone	$(C_6H_9NO)_x$	~10 000	Sigma-Aldrich
Potassium chromate	KCr_2O_4	194.2	Sigma-Aldrich
Pyrrole	C_4H_5N	67.1	Sigma-Aldrich
Rhodium chloride trihydrate	$RhCl_3 \cdot 3H_2O$	209.3	Sigma-Aldrich
Sodium chloride	$NaCl$	58.4	Merck
Triethanolamine	$C_6H_{15}NO_3$	149.2	VWR
Trithiocyanuric acid	$C_3H_3N_3S_3$	177.3	Abcr
Thiobarbituric acid	$C_4H_4N_2O_2S$	144.2	Sigma Aldrich

3.3 Syntheses

3.3.1 Synthesis of g-C$_3$N$_4$

15 g of melamine were suspended in 75 ml water, the water was evaporated and the sample calcined for 4 h at 550 °C under nitrogen atmosphere. The heating rate was 2.3 °C min^{-1} and the sample was allowed to cool down to room temperature under nitrogen atmosphere in the oven. The obtained product was pestled and characterized.

3.3.2 Synthesis of Copolymerized g-C$_3$N$_4$

15 g melamine or dicyandiamide as precursor and a defined amount of comonomers were mixed. In order to get a finely dispersed mixture, the compounds were dissolved in hot water (temperature was varying). The water was evaporated and the mixture was pestled. Table 3-3 and Table 3-3b give detailed information about the used precursors and comonomers. The received sample was calcined for 4 h at 550 °C under nitrogen atmosphere. The heating rate was 2.3 °C min^{-1} and the sample was allowed to cool down to room temperature under nitrogen atmosphere in the oven. The obtained product was pestled and characterized.

3.3.3 Dye Modification of g-C$_3$N$_4$ by Chemical Synthesis

The dye modification of differently copolymerized graphitic carbon nitride was done by DIC (*N,N'*-Diisopropylcarbodiimide) and DMAP (4-Dimethylaminopyridine) as coupling agents. In order to check the influence of these agents, the synthesis was also conducted without the reagents. At first, the carbon nitrides were activated by heating them in concentrated hydrochloric acid under reflux. The semiconductor was filtered and washed with 100 ml water and 50 ml acetonitrile. Following a general method, 15 mg of erythrosine B (0.011 mmol, 1 eq.), 21 mg 4-dimethylaminopyridine (0.057 mmol, 5 eq.) and 24 µl *N,N'*-diisopropylcarbodiimide (0.057 mmol, 5 eq.) were added to 150 ml of acetonitrile. Hereinafter, the reaction mixture was equilibrated at room temperature. After 1 h, the addition of 750 mg varying graphitic carbon nitrides followed. After 3 h the reaction mixture was filtered and washed with 100 ml acetonitrile. Used compounds for dye immobilization are: g-C$_3$N$_4$ (MA), g-CN$_{BA\_0.5}$ (MA), g-CN$_{BA\_1.0}$ (MA), g-CN$_{BA\_2.1}$ (MA), g-CN$_{BA\_5.0}$ (MA), g-CN$_{BA\_8.2}$ (MA), g-C$_3$N$_4$ (DCA), g-CN$_{BA\_1.0}$ (DCA) and g-CN$_{BA\_5.0}$ (DCA).

Table 3-3: *Composition of different monomers for graphitic carbon nitride synthesis*

Compound	Precursor	Comonomer	Mass of Comonomer / g	Relative* $\chi_{comonomer}$ / mol-%
g-CN$_{BA\_0.5}$ (MA)	Melamine	Barbituric acid	0.25	0.5
g-CN$_{BA\_1}$ (MA)	Melamine	Barbituric acid	0.50	1.0
g-CN$_{BA\_2.1}$ (MA)	Melamine	Barbituric acid	1.00	2.1
g-CN$_{BA\_5}$ (MA)	Melamine	Barbituric acid	2.50	5.0
g-CN$_{BA\_8.2}$ (MA)	Melamine	Barbituric acid	5.00	8.2
g-CN$_{BA\_1}$ (DCA)	Dicyandiamide	Barbituric acid	0.50	1.0
g-CN$_{BA\_5}$ (DCA)	Dicyandiamide	Barbituric acid	2.50	5.0
g-CN$_{CA\_1}$ (MA)	Melamine	Cyanuric acid	0.50	1.0
g-CN$_{CA\_5}$ (MA)	Melamine	Cyanuric acid	2.52	5.0
g-CN$_{CA\_1}$ (DCA)	Dicyandiamide	Cyanuric acid	0.50	1.0
g-CN$_{CA\_5}$ (DCA)	Dicyandiamide	Cyanuric acid	2.52	5.0
g-CN$_{2MP\_1}$ (MA)	Melamine	2-mercaptopyrimidine	0.56	1.0
g-CN$_{2MP\_5}$ (MA)	Melamine	2-mercaptopyrimidine	2.78	5.0
g-CN$_{2MP\_1}$ (DCA)	Dicyandiamide	2-mercaptopyrimidine	0.56	1.0
g-CN$_{2MP\_5}$ (DCA)	Dicyandiamide	2-mercaptopyrimidine	2.78	5.0

*   Please note that the absolute ratio of the two comonomers varies since melamine and dicyandiamide have   different molecular masses. Therefore, the relative $\chi_{comonomer}$ is calculated to the amount of nitrogen atoms, which is constant for the same mass of dicyandiamide and melamine.

Table 3-3 (contd): Composition of different monomers for graphitic carbon nitride synthesis

Compound	Precursor	Comonomer	Mass of Comonomer / g	Relative* $\chi_{comonomer}$ / mol-%
g-CN$_{2TB\_1}$ (MA)	Melamine	2-thiobarbituric acid	0.56	1.0
g-CN$_{2TB\_5}$ (MA)	Melamine	2-thiobarbituric acid	2.81	5.0
g-CN$_{2TB\_1}$ (DCA)	Dicyandi-amide	2-thiobarbituric acid	0.56	1.0
g-CN$_{2TB\_5}$ (DCA)	Dicyandi-amide	2-thiobarbituric acid	2.81	5.0
g-CN$_{TTC\_1}$ (DCA)	Melamine	Trithiocyanuric acid	0.69	1.0
g-CN$_{TTC\_5}$ (MA)	Melamine	Trithiocyanuric acid	3.46	5.0
g-CN$_{TTC\_1}$ (DCA)	Dicyandi-amide	Trithiocyanuric acid	0.69	1.0
g-CN$_{TTC\_5}$ (DCA)	Dicyandi-amide	Trithiocyanuric acid	3.46	5.0
g-CN$_{6MI\_1}$ (MA)	Melamine	6-Methylisocytosine	0.63	1.0
g-CN$_{6MI\_5}$ (MA)	Melamine	6-Methylisocytosine	3.15	5.0
g-CN$_{6MI\_1}$ (DCA)	Dicyandi-amide	6-Methylisocytosine	0.63	1.0
g-CN$_{6MI\_5}$ (DCA)	Dicyandi-amide	6-Methylisocytosine	3.15	5.0

*   Please note that the absolute ratio of the two comonomers varies since melamine and dicyan-
       diamide have   different molecular masses. Therefore, the relative $\chi_{comonomer}$ is calculated to the
       amount of nitrogen atoms, which is constant for the same mass of dicyandiamide and mela-
       mine.

3.3.4   Modification of g-C$_3$N$_4$ by Electron Beam

Another approach to modify various polymers, used to immobilize certain small molecules on the surface, was developed by the Leibniz Institut for Oberflächenmodifizier-
ung e.V.[76] The modifications are done by the Elektronen-Laborbeschleu-
nigungsanlage (ELSA). While using 200 kV for electron acceleration, the absorbed

energy dose can be verified by the contact time of the sample with the electron beam. Herein, the absorbed dose was kept constant at 200 kGy. In order to test the efficiency of the electron immobilizing of erythrosine B on the surface, modification was also carried without electron beam treatment according to the procedure as follows. Thus, the dye was just immobilized by wetness impregnation. 500 mg g-C_3N_4 (MA) were spread out on a glass plate (diameter: 4.6 cm). 10 mg dye was added to certain samples. The prepared catalysts were suspended in 1.5 ml water and treated with the electron beam three times. Afterwards the samples were washed in 50 ml water to remove the adsorbed dye and filtrated. The procedure is summarized in Table 3-4.

Table 3-4: *Details for electron beam treatment of g-C_3N_4 (MA)*

Compound	Activation	Dye	Treatment
$_eg$-C_3N_4	not activated	-	electron beam
$_{e\text{-}act}g$-C_3N_4	activated	-	electron beam
Dye$_e$@g-C_3N_4	not activated	erythrosine B	electron beam
Dye$_{e\text{-}act}g$-C_3N_4	activated	erythrosine B	electron beam
g-C_3N_4	not activated	-	-
$_{act}g$-C_3N_4	activated	-	-
Dye$_{Wl}$@g-C_3N_4	not activated	erythrosine B	Wetness impregnation
Dye$_{Wl\text{-}act}g$-C_3N_4	activated	erythrosine B	Wetness impregnation

3.3.5 Synthesis of Pt and Rh Nanoparticles

The synthesis of the NPs was done according to *Teranishi et al.*[33] In a typical approach, 1 ml of a 40 mM aqueous solution of H_2PtCl_6 or $RhCl_3$ was evaporated in a flask, the flask was purged with nitrogen and 159 mg PVP, 10 ml ethylene glycol and 45 μl HNO_3 were added. The reaction mixture was heated for 16 h at 120 °C. Hereinafter, the resulting solution of nanoparticles was purified by adding 2 ml isopropanol and extracting the organic impurities three times with 5 ml hexane. The residual ethylene glycol was removed under reduced pressure. The stabilized nanoparticles were resolved in 50 ml absolute ethanol.

3.3.6 Deposition of PVP Stabilized Pt and Rh Nanoparticle

300 mg of g-C_3N_4 were added to the resulting nanoparticle solution from section 3.3.5. The noble metal nanoparticles which were coated by polyvinylpyrrolidone were supposed to be immobilized on the surface by electrostatic adsorption. After 20 h stirring at room temperature, the modified catalyst was filtered, washed with ethanol and cal-

cined at 400 °C for 3 h to remove PVP. In order to gain more modified catalyst, the batch could be enlarged.

3.3.7 Rh and Pt Nanoparticle Deposition by Visible Light

Though it was the aim to get 1.5 wt.-% Pt^0 on g-C_3N_4, the molar amount was kept constant for comparability for both used noble metals, Rh and Pt. A relationship of used noble metal salt and obtained elemental noble metal was found previously. Therefore, 1.0 g graphitic carbon nitride (derived from melamine) was suspended in 100 ml water with a subsequent addition of 75.5 mg H_2PtCl_6 · 6 H_2O or 38.4 mg $RhCl_3$ · 3 H_2O, respectively. The reaction was carried out under nitrogen atmosphere. A 250 W and 220 V light bulb by *MLW Labortechnik Ilmenau* was used for the photo reduction of the metals. Visible light was applied with a wavelength above 300 nm using amorphous glass as a cut-off filter. Distance to samples was kept constant. The solution was irradiated for 4 h while the solution was heated under reflux.

3.3.8 Cr_2O_3 Nanoparticle Deposition by Visible Light

Cr_2O_3 nanoparticle deposition or nanoparticle shell growth, respectively, by Light was done. Therefore catalysts which were synthesized according to section 3.3.1 and 3.3.6 were modified as follows. In a water cooled flask, 1 g of the catalyst was suspended in 100 ml H_2O and 3.88 g K_2CrO_4. The reaction was conducted under nitrogen. For photodeposition of Cr_2O_3 250 W and 220 V light bulb by *MLW Labortechnik Ilmenau* was used. Visible light was applied with a wavelength above 300 nm using amorphous glass as a cut-off filter. Distance to samples was kept constant. The suspension was irradiated for 4 h at room temperature.

3.3.9 Synthesis of Co_3O_4 Nanoparticle

0.5 g of $Co(OAc)_2$ · 4 H_2O or 0.63 g of $CoCl_2$ · 6 H_2O were dissolved in an a mixture of ethanol, water and ammonia.[77] Detailed information is given in Table 3-5. After homogenizing, the Co^{2+} solution was transferred in to a 50 ml autoclave, sealed and incubated for 3 h at 150 °C. After this, the autoclave cooled down naturally within the oven. The resulting suspension of Co_3O_4 nanoparticles was centrifuged at 1000 rpm for 20 min, the supernatant was discarded and the nanoparticles were dispersed in 30 ml water. These steps were repeated three times. Afterwards, the nanoparticles were dried at 60 °C for 4 h. Attempts Co_3O_4 NP I through to Co_3O_4 NP III did not yield any nanoparticles.

Table 3-5: Detailed synthesis parameters of Co_3O_4 nanoparticles

Compund	Co^{2+} Salt	$V_{Ethanol}$ / ml	V_{Water} / ml	$V_{Ammonia\ (25\ \%)}$ / ml
Attempt Co_3O_4 NP I	$CoCl_2 \cdot 6\ H_2O$	25	0.0	2.5
Attempt Co_3O_4 NP II	$CoCl_2 \cdot 6\ H_2O$	23	2.0	2.5
Attempt Co_3O_4 NP III	$CoCl_2 \cdot 6\ H_2O$	15	10	2.5
s-Co_3O_4 NP	$Co(OAc)_2 \cdot 4\ H_2O$	25	0.0	2.5
m-Co_3O_4 NP	$Co(OAc)_2 \cdot 4\ H_2O$	23	2.0	2.5
l-Co_3O_4 NP	$Co(OAc)_2 \cdot 4\ H_2O$	15	10	2.5

3.3.10 Modification of g-C_3N_4 with Co_3O_4 Nanoparticle

To accomplish the modification of g-C_3N_4 with cobalt oxide, two separate synthesis routes were done. Therefore, the material synthesized accordingly to 3.3.9 was supposed to be deposited on the basic catalyst g-C_3N_4 (MA) in two different ways.

In the first experiment, the different cobalt oxides were suspended water in which 2.22 g of melamine were dissolved. This was followed by the evaporation of water. This incipient wetness impregnation method aimed for a well-dispersed cobalt oxide and carbon nitride mixture, hence, the water was completely evaporated. For these synthesis routes either 10 mg or 1 mg of Co_3O_4 NPs were used, which should result in the material being doped with 1.0% and 0.1% Co_3O_4, respectively. The resulting mixture was calcinced according to the procedure described in section 3.3.1.

The second experiment was a classical electrostatic adsorption deposition of nanoparticle. Therefore,10 mg of the Co_3O_4 particle were suspended as well as 1.0 g graphitic carbon nitride in 100 ml water and stirred for 24 h, followed by the filtration and drying of the resulted powder.

3.3.11 Polypyrrole Modification of g-C_3N_4

Because a possible connection of noble metal nanoparticle deposition on graphitic carbon nitride and dye sensitizing of the same material should be confirmed, a new method was tried to accomplishing both. Therefore, 113 mg $H_2PtCl_6 \cdot 3\ H_2O$, 130 mg NaCl, and 26 μl pyrrole were added to 100 ml H_2O, followed by the addition of 1.0 g g-C_3N_4 derived by melamine. The resulting suspension was stirred for 4 days. The dark black material (Pt_{Ppy}@g-C_3N_4 pure) was washed and filtrated. In additional experiments, it was tested to obtain a dye modified catalytic material. These experiments were conducted with 6.6 mg and 66 mg of erythrosine B, respectively, were added with the pyrrole and the other compounds (Pt_{Ppy}@g-C_3N_4 Dye I and Pt_{Ppy}@g-C_3N_4 Dye II).

3.4 Characterization

3.4.1 Bulk Characterization

In order to get an overview on the structure of g-C_3N_4, matrix assisted laser desorption ionization mass spectrometry (Maldi-MS) was performed. During such a process it is possible to ionize and analyze molecules and larger fragments with masses up to 20.000 Da. Daltonics spectrometer by *Bruker* was used.

Powder X-ray diffraction (XRD) was used to identify the graphitic carbon nitride samples, because the material shows typical signals in a XRD pattern. For all materials, the chosen angle (2θ) was between 4 and 85°. XRD experiments were conducted at the institute for mineralogy, crystallography and material science of the Universität Leipzig. The XRD-7 (Seifert) was used for these experiments. Copper Kα radiation with a wavelength λ of 1.541 nm was used.

3.4.2 Computational Details

For density functional theory calculations the program *Orca* (program version 2.9.1) was used, which is package for ab initio, DFT and semiempirical electronic structure calculation. It was developed by the Max Planck Institute for Bioinorganic Chemistry, Mühlheim. In order to get proper results for HOMO and LUMO calculations a geometry optimization with the Ahlrichs bases def2-TZVP and the BLYP functional was done.[78] Subsequently, the coordinates were used for a single-point calculation using the same bases and the B3LYP functional. The later has the advantage of being a hybrid functional using exact energies derived from a Hartree-Fock calculation.[79] A Hartee energy of 10^{-6} was the convergence criterion was in both cases. Nevertheless, an empirical van-der-Waals correlation was done.[80]

3.4.3 Elemental Analysis

The amounts of carbon, nitrogen and sulfur were determined by classical approaches by combustion analytic. The samples are weighted and burned in pure oxygen. The resulting gases (CO_2, NO_x and SO_2) – NO_x gets reduced to N_2 – were analyzed via gas chromatographically and with a TCD. The experiments defining the carbon, nitrogen and sulfur content were conducted with Vario IL (*Elementaranalyse Systeme Hanau*).

Because nanoparticle synthesis was conducted, elemental analysis of platinum, rhodium, chromium and cobalt had to be done. This was accomplished by means of inductive coupled plasma optical emission spectroscopy (ICP-OES). 50 mg of the samples were solved by microwave treatment 6 ml HCl (35 %) and 6 ml HNO_3 (69 %) (Rotipuran supra, purchased from Roth). *Perkin and Elmers* Optima 8000 was used for ICP-OES measurements. This method uses a plasma torch to excite atoms and detects the emitted light during that process. It is possible to analyze the light with a UV camera sphere in connection with a charge coupled detector. Because every element has a characteristic spectrum and the amount of emitted light correlates to the amount of used material used, it is possible to analyze the samples qualitatively and quantitatively.

3.4.4 Nanoparticle Tracking Analysis

Nanoparticle tracking analysis is a tool to determine the size of macromolecules like proteins or stabilized nanoparticles in solution, because the Brownian motion of particles in solution just depends on viscosity, temperature and particle size.[81] Hence, it is possible to determine a size distribution profile for nanoparticles which have a diameter between 10 to 1000 nm. By using an ultra-microscope under laser illumination the Stokes-Einstein equation can be used to evaluate the hydrodynamic radius which is equivalent to the sphere radius. These measurements were done with the LM10 *NanoSight*.

3.4.5 Surface Characterization

The nitrogen adsorption was done using an ASAP 2010 (*Micromeritics Instruments*). The principle of measurement is based on adsorption and desorption of nitrogen. In this project the samples were activated at 150 °C. Therefore, the investigated material is cooled down close to the boiling point of nitrogen. The evacuated samples are treated with nitrogen at atmospheric pressure and about 78 °K. Hereinafter, the system is again evacuated. From the so obtained nitrogen isotherms Brunauer, Emmet and Teller developed an analysis which uses the hysteresis curves to determine the specific surface.

The characterization of a semiconductor via UV/Vis spectroscopy is fundamental, diffuse reflectance UV/Vis spectroscopy was performed. In these experiments an incident ray is reflected by a solid surface in many angles instead of passing a sample. The reflected light is collected by a universal reflectance sphere for all wavelengths, which allows to record a wavelength depended UV/Vis. A luminescence will appear in all direction lying in the half-space adjacent to the surface, the so called Lambertian reflectance. All samples were measured between 200 and 800 nm. The split width was 2 nm.

Transmission Electron Microscopy (TEM) is able identify nano-sized structures and objects on solid surfaces. The principle of the measurement is based on accelerated electrons which function as the testing probe and can be spatially resolved by analyzing the Rutherford scattering. While an evacuated sample is hit by a passing electron, unalike compounds interact differently with the emitted electrons. Contrasting microscopic images can be recorded, when a (nano-sized) object interacts much more with the probe than the support, e.g. noble metals on oxides. Thus, depth profiles can be recorded in the same manner. Samples were prepared by sonicating 5 mg of the material in 0.5 ml acetone solution for 10 min. Afterwards 5 µl of this solution were casted on a copper grid with carbon support film which has been pretreated by glow discharge. The liquid was evaporated by drying. The prepared sample was transferred to the TEM (Jem-2100 by *Jeol*). The TEM was operated at an acceleration voltage of 200 kV. All images were recorded digitally.

3.4.6 Thermal Characterization

In order to define the processes happening during polymerization, thermogravimetrical analysis (TG) and differential thermal analysis (DTA) coupled with mass spectrometry were conducted. In this manner, melamine and dicyandiamide and mixtures with barbituric acid were analyzed. The principle of measurement includes heating the sample with a defined heating rate while the loss of weight and temperature difference to an inert reference material. Therefore 50 mg of the samples were weighed in. A temperature ramp of 2 °K min^{-1} was used. The used equipment was STA 09 from *Netzsch*. In order to define the degassing compounds during the heating process, an ion impact mass spectrometer with a capillary system was used.

Furthermore, temperature programmed reduction (TPR) was used to characterize the effect of deposited nanoparticle promoters for hydrogen reaction. Therefore, platinum modified as well as pure g-C_3N_4 were investigated using this method. The samples were pretreated at 150 °C for 1 h, hereinafter, treated with a 5.18 vol.-% hydrogen argon mixture as gas flow and heated with a rate of 2 °K/min. The consumption of hydrogen was measured by a thermal conductivity detector in order to survey the reduction of the material. The measurement was performed at the AMI-100. The results allowed a qualitative characterization of the samples.

3.5 Photocatalytic Hydrogen Evolution from Water

The photocatalytic hydrogen evolution experiments were conducted in a self-made apparatus which is schematically depicted in Figure 3-1. For the experiment 250 mg catalyst were suspended in the ultra-sonic bath for at least half a minute in 360 ml H_2O and 40 ml MeOH. MeOH had to be used for a sufficient hydrogen evolution. It functions as a sacrificial agent which gets oxidized instead of water. In preliminary experiments the MeOH amount was changed but no significant change of the hydrogen evolution rate was observed. Additionally for some catalytic experiments, 3.32 mg $H_2PtCl_6 \cdot H_2O$ were added and then photoreduced to Pt^0 resulting in a noble metal deposition ω_{Pt} of 0.5 wt.-% Pt on the graphitic carbon nitride. This was the standard treatment. For the testing of platinum and rhodium modified g-C_3N_4 no Pt salts were added. The suspended powder was spinning in circles. Furthermore, the reaction chamber was flushed with nitrogen for 3 min.

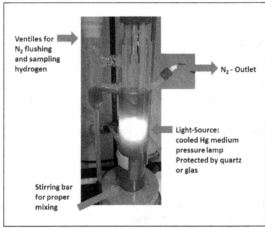

Figure 3-1: *Partly schematic picture of the vessel system for photocatalytic water splitting with suspended powder*

A UV Hg medium pressure lamp (150 W) by *Heraeus* was used to irradiate the samples. Therefore, the UV lamp was protected by an immersion quartz or glass which was cooled with water to keep the lamp and the aqueous medium at 20 °C. The spectrum of the lamp is shown in the appendix (Figure 0-1). In order to guarantee a constant level of light emission from the lamp, it was allowed to warm up for 15 min before the first irradiation experiment and in between each following experiment as well. The suspended samples for the catalytic water splitting by UV/Vis light were irradiated for 60 min. Afterwards, the gas was pumped in circles around the GC system. After 5 s a 250 µl gas sample were injected into the GC system. Nitrogen was used as carrier gas. The sampled gases were separated on a column for gas chromatography (*Haysep*) at 50 °C. The temperature at the TCD detector was 250 °C. The analysis of the evolved hydrogen was done qualitatively by the retention time and quantitatively by the peak area of a TCD detector. For quantitative analysis of the hydrogen amount, the system was preliminary calibrated using a mass flow controller to regulate the hydrogen stream. Hence, the peak area-hydrogen amount was determined.   Each measurement was done twice. The hydrogen evolution rate was ascertained by the measured hydrogen amount over the time of light irradiation. The reaction rate can be obtained in this manner, because the hydrogen evolution is a zero-order reaction.

For dye-sensitized graphitic carbon nitride amorphous glass was chosen to cut off the irradiation spectrum of the lamp below a wavelength of about 300 nm. For the other carbon nitrides quartz was used which allows light to pass till a wavelength of about 200 nm.

4 Results and Discussion

4.1 Material Characterization and Improvement

4.1.1 Pure g-C_3N_4

For a modification of graphitic carbon nitrides there are two starting compounds which are suitable for a sufficient synthesis of the semiconductor. One aim of the work was to analyze and improve the graphitic carbon nitride described and investigated by several groups so far. Pure g-C_3N_4 was synthesized according to well-known procedures as described in section 3.3.1 from melamine (MA) or dicyandiamide (DCA) and differences in physical and chemical properties as well as their catalytic activity were studied. XRD pattern and UV/Vis spectra are depicted in Figure 4-1 and Figure 4-2 indicate that the catalysts derived from two different precursors do not vary strongly.

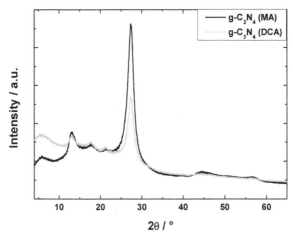

Figure 4-1: XRD pattern from graphitic carbon nitride derived from melamine (MA) and dicyandiamide(DCA)

XRD and UV/Vis characterizations prove the successful synthesis of g-C_3N_4.[49] Typically, a signal at 13°, which is caused by distances between two unit meshes of the g-C_3N_4, can be found. Furthermore, the signal at 27° can be assigned to the distance between the layers of heptazine sheets. Since graphitic carbon nitride is amorphously or finely structured, the signals are very broad. Because the structure has never been entirely elucidated, small signal at 45° and 55° and possible covered signal at 26° might be a contribution from graphite as a side product.[82] Nevertheless, graphite XRD signals would slightly differ from XRD signals presented herein. Though, this is strong hint how close the structures of graphite and graphitic carbon nitride really are.

Still, melamine and dicyandiamide derived graphitic carbon nitrides differ slightly in their XRD patterns. Interestingly, the peak at 27° is considerably more pronounced for g-C$_3$N$_4$ derived from melamine. Thereby, it can be concluded, that the interaction between the layer-stacking is stronger for melamine derived carbon nitride. Thus, g-C$_3$N$_4$ (MA) has a better ordered than g-C$_3$N$_4$ (DCA).[47] This might result in different possibilities for molecules to interact with the solid catalyst. Unfortunately, the exact structure of graphitic carbon nitrides has not been elucidated.[53]

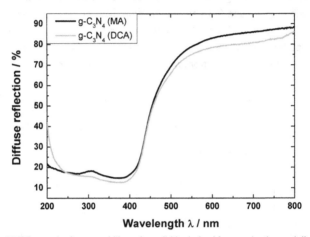

Figure 4-2: *UV/Vis spectra from graphitic carbon nitride derived from melamine and dicyandiamide*

Furthermore, the diffuse reflection UV/Vis spectrum shows similar behavior of both compounds. The absorption edge, which describes the point on the curves of the highest slope, is around 435 nm in both cases. The semiconductor is most commonly described as (pale) yellow; hence, it can still absorb visible light. Considering the efficiency of this material as a photocatalyst, it is noticeable that, regarding a range from 200 to 800 nm, just about 45 % of the light will be converted into non radiation energy. As mentioned before, a wavelength of 800 nm could theoretically still suffice for water splitting.

To measure of the quantity of light absorbance, integration of diffuse reflection UV/Vis spectra were carried out. This will help to compare dye sensitizing and copolymerization processes later on (sections 4.1.2 and 4.2). The integration of the spectra was carried out with the program Origin 8.0. The integration limits were 200 nm and 800 nm. The integrated area of the spectrum was 328 nm for g-C$_3$N$_4$ (MA) and 312 nm for g-C$_3$N$_4$ (DCA). The so derived values illustrate how much irradiation is absorbed by the used carbon nitrides. Low values describe a high light absorption. In the following, the integrated area of the spectra will be referred as totally reflected light δ. Thus, a totally black material would have an integrated area of 0 nm.

Table 4-1: Elemental analysis regarding pure graphitic carbon nitrides

Compound	χ_C / mol-%	χ_N / mol-%	C/N Ratio
g-C$_3$N$_4$ (MA)	39.93	60.07	0.66
g-C$_3$N$_4$ (DCA)	40.33	59.67	0.67

So far, the compounds are not distinguishable by UV/Vis analysis. Still, there are minor differences at both sides of the absorption edge. The small variances from the UV/Vis spectra imply that two graphitic carbon nitrides might differ in their catalytic performance. Nevertheless, these results following previous results.[51]

Table 4-2: Expectations for elemental analysis depending on the structure of g-C$_3$N$_4$

Structure	Elemental Structure Motif	№ of C Atoms	№ of N Atoms	C/N Ratio
2 dimensional connected	Unit Mesh	6	8	0.75
1 dimensional connected	Monomeric Unit	6	9	0.67

Since the investigated graphitic carbon nitrides were almost invariant in their physical properties, it is not surprising that their composition was equally identical. Just minor variations were found (Table 4-1). Although it was imaginable that the structure might be modified by altering the precursor, no such influence was found. This will be even more interesting in the discussion of DTA results (section 4.1.2). Because the ratio of carbon to nitrogen (C/N) should be depending on the polymeric organization, structure motifs might be assigned by elemental analysis. As explained in Table 4-2, due to another organization, one dimensional connected heptazine polymers should have C/N ratio of 0.67, while sheets of heptazines have a ratio of 0.75.

It could be concluded that the graphitic carbons nitrides investigated herein are one dimensionally structured. On the other hand, these ratios, which were lower than expected for a two-dimensional structure, might arise from a higher nitrogen loss during preparation of a two dimensionally structured g-C$_3$N$_4$.[49] Since more detailed investigations concerning this topic were done, this might be just a minor indication.[53] Nevertheless, it is uncertain whether the material should be called g-C$_2$N$_3$ instead of g-

C_3N_4. Despite this, established nomenclature was not changed. The results, which have been presented so far, are just for the characterization of the basic material. Because the structure of pure graphitic carbon nitride should be elucidated a little further, matrix assisted laser desorption ionization mass spectrometry (Maldi-MS) was performed (Figure 4-3). Surprisingly, this technique has not been used so far to investigate the structure of g-C_3N_4.

Three groups of signals were identified, which can assigned to the three different heptazine ring systems: a monomeric unit (blue), a dimer of heptazine (red) and a trimer of heptazine (green).Though there is rather a wide range of products than one main peak, the strongest signals can be assigned to the main species for each heptazine system. Due to the soft ionization technique (Maldi), compounds will be single charged. Thus the m/z ratios can be assigned to the masses of the compounds in Dalton (Da).

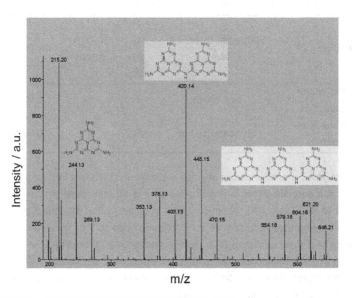

Figure 4-3: *MALDI mass spectrum from pure g-C_3N_4 (MA)*

The monomer, dimer and trimer of heptazine were identified according to the molecular weight of heptazine as the smallest detected unit. The molecular mass of heptazine (M_1) is 218 Da. $[M_1+H]^+$ can be found with 219 Da next to 215 Da (matrix effect). Furthermore, are there are peaks according to $[M_1+25+H]^+$ and $[M_1+50+H]^+$ (244 Da, 269 Da). In this context, there might be two explanations for this phenomenon. Clearly, signals from $[M_1+25+H]^+$ have to rise when a terminal –NH_2 groups of heptazine (M_1) are replaced with –NH-CN groups. On the one hand, that might come from collision induced dissociation of a dimer during the mass spectrometric performance. A dimer of heptazine rings would have dissociated into a larger and a smaller fragment during the ionization. On the other hand, the whole detected heptazine derivative might be modified during synthesis.

The same consideration can be done when it comes down to interpret the results of dimers (red) and trimers (green) of heptazine (Figure 4-3). A heptazine dimer has a molecular mass M_2 of 419 Da which can be found as $[M_2+H]^+$ at 420 Da. A loss of 17 Da is surely identified as leaving ammonia, while the losses of 50 Da are supposed to be the process described above for $[M_1]$. Herein, instead of addition of CN and a proton and a nitrile group are lost. Almost alike, spectrum can be seen in the range of 554 to 646 m/z. Although it might be discussed whether a trimer is made of a polymeric chain ($M_{3P} \triangleq 603$ Da) or builds a 2-dimensional unit mesh ($M_{3M} \triangleq 620$ Da). Unfortunately, the mass difference of theoretical structure motifs is exactly 17 Da. Hence, m/z $[M_{3P}+H]^+$ (604) and $[M_{3M}+H]^+$ (620) might be found in the spectrum as a sign for both species or a 2-dimensional motif with leaving of ammonia.

Overall speaking, the Maldi mass spectrum raises a question concerning the structure of graphitic carbon nitride. The results from the Maldi-MS experiments are unique in the structure elucidation of graphitic carbon nitrides.[51] Though matrix assisted laser desorption ionization is able to vaporize large molecules (e.g. peptides or proteins) with masses up to 20,000 Da, just monomers, dimers or trimers were found to be ionized and analyzed by this method when it was applied on g-C_3N_4. Furthermore, masses of pure heptazine monomers, dimers or trimmers were found with no lack of hydrogen indicating that these molecules were not created by collision induced dissociation, but existed free within the matrix. This could imply that graphitic carbon nitride has at least partially free heptazine rings. High molecular compounds might not be detected due to hindered evaporation or ionization.

Furthermore, TG and DTA measurements of the pure graphitic carbon nitride in air were carried out (Appendix: Figure 0-2). As it is expected for these kinds of compounds, 1.9 wt.-% of adsorbed water were released up to 445°C followed by the decomposition of graphitic carbon nitride between 510 and 600 °C. Since preparation of g-C_3N_4 is conducted between those temperatures, these temperatures were expected.[54]

Table 4-3: Hydrogen evolution rate of pure g-C_3N_4

Compound	Hydrogen Evolution / $\mu mol\ h^{-1}$
g-C_3N_4 (MA)	182
g-C_3N_4 (DCA)	140

Nevertheless, the aim was to improve graphitic carbon nitride for light-driven hydrogen evolution. Because every sample setup is reported to be slightly different in literature, it is necessary to measure and compare the basic catalytic material derived from both precursors. It is shown in Table 4-3, that g-C$_3$N$_4$ (MA) has a significantly higher activity towards photocatalytic water splitting than g-C$_3$N$_4$ (DCA). Their values of hydrogen evolution will be taken in account for the comparison of the efficiency of the improvement of the catalytic activity. Because the apparatus setup for photocatalytic hydrogen evolution from water was reported to be always different, there are no standard values and the results presented in this thesis will be normalized to these values in Table 4-3.[65]

4.1.2 Copolymerization of g-C$_3$N$_4$

As it was already mentioned, graphitic carbon nitrides can be modified by copolymerization of dicyandiamide or melamine with other comonomers. Barbituric acid was chosen by *Zhang* and coworkers.[70] They introduced barbituric acid to increase the carbon amount of carbon in g-C$_3$N$_4$. This led to a red-shifted absorption edge. Since this method was successful, copolymerization could be a good method for the introduction of other heteroatoms.[65]

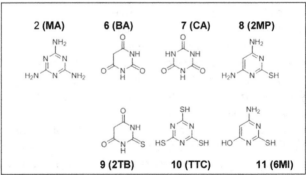

Figure 4-4: *Structure of comonomers used to improve the basic catalytic material*

Therefore, the aim of this present thesis is the investigation of the sufficiency of copolymerization reactions in order to explore the potential of the modification and development of g-C$_3$N$_4$ with sulfur in a single one step reaction. This method would be much more beneficial compared to sulfur-modifications already reported.[72] Unsuccessful examinations were reported testing the possibility to derive g-C$_3$N$_4$ from diaminobenzene. Thus, melamine and dicyandiamide were mixed with the compounds depicted in Figure 4-4 (**6** to **11**) and calcined under nitrogen.

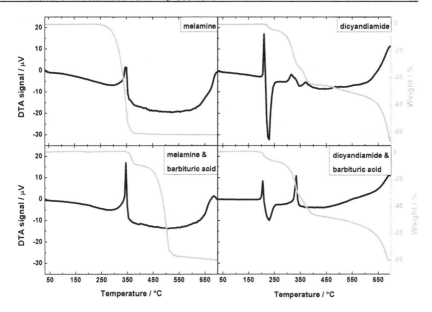

Figure 4-5:   *DTA (black) and TG (grey) of several precursors of carbon nitrides: copolymerization analysis was done with mixtures according to syntheses of g-CN$_{BA\_1}$ (MA) and g-CN$_{BA\_1}$ (DCA).*

In order investigate the influence of the used precursors for a copolymerization process, different thermal analyses were carried out at first (Figure 4-5). Although there have been formation investigation towards the reaction from melamine to graphitic carbon nitride, the copolymerization process of melamine and barbituric acid has been just a proposed mechanism.[49]

DTA and TG were exhibited for pure melamine, dicyandiamide and mixtures of both precursors with barbituric acid (**6**). As already mentioned, the latter was reported to lead to a red-shift in UV/Vis spectrum of the resulting g-C$_3$N$_4$.[70] Obviously, the polymerization process for both materials differ, even then, if it is polymerized with the same copolymer. Exothermic reactions can be assigned by a positive signal. The polymerization of melamine shows only one large exothermic peak at about 350 °C (Figure 4-5). This is supposed to be a superimposed exothermic polymerization, crystallization or decomposition over an endothermic degassing process. The leaving compounds were identified by mass spectrometry to be ammonia (fragment 17 ≙ NH$_2$) and carbon dioxide (fragment 44 ≙ CO$_2$). The mass spectra can be found in the appendix (Figure *0-4* and Figure 0-5).

A more straight forward interpretation is possible, if the reaction of melamine at 350 °C is compared with the first reactions of dicyandiamide at 350 °C (Figure 4-5). Comparable reactions have to be involved, due to the proposed reaction pathway and resulting products. The first reactions of dicyandiamide between 200 °C and 250 °C might be explained by reorganization with a starting polymerization process, in which melamine or other higher molecular compounds are formed. Although the polymerization begins at 350 °C, there are fragments from ammonia and carbon dioxide at the reorganization in lower temperature regions. While analyzing precursor and copolymer reaction, the more exothermic peak at 350 °C for both precursors should be discussed.

Table 4-4: Elemental analysis regarding g-CN_{BA_X}

Compound	χ_C / mol-%	χ_N / mol-%	C/N Ratio
g-$CN_{BA\_0.5}$ (MA)	40.0	60.0	0.67
g-CN_{BA_1} (MA)	40.3	59.7	0.68
g-$CN_{BA\_2.1}$ (MA)	41.1	58.9	0.70
g-CN_{BA_5} (MA)	40.7	59.3	0.69
g-$CN_{BA\_8.2}$ (MA)	45.9	54.1	0.85
g-CN_{BA_1} (DCA)	40.6	59.4	0.69
g-CN_{BA_5} (DCA)	42.6	57.4	0.74

Since the signals at 350 °C might be caused product formation and degassing (Figure 4-5), a higher endothermic reaction implies that a superimposed reaction is less exothermic than the reaction without copolymer. Therefore, the formation process with copolymers might occur not as completely as with pure precursors. In general, thermal analysis reveals the nature of the copolymerization process, which is disturbed by copolymers. This has not been shown yet.[49]

To study the influence of the ratio, six different mixtures of melamine and barbituric acid were investigated. The product was studied with elemental analysis (Table 4-4). Elemental analysis gives a quick overview on the performed modifications by copolymerization.

The content of carbon to nitrogen varies with altered copolymer ratios. The C/N ratio increases from 0.67, which is comparable to pure g-C_3N_4, to 0.85. Regarding the structure of the catalytic material as a chain, the latter ratio implies that – on average – 1.7 nitrogen atoms were replaced with carbon atoms within one heptazine ring. Barbituric acid contains one carbon instead of a nitrogen atom. A monomeric unit should have 9 nitrogen atoms and about 25 mol-% of barbituric acid were used. But for a full conversion of barbituric acid into the carbon nitride structure 2.2 nitrogen atoms are supposed to be replaced. It can be assumed that for that ratio of barbituric acid just 77 % of the copolymer is introduced to the structure.

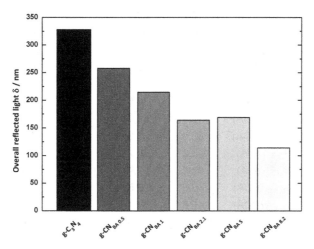

Figure 4-6: *Overall reflected light from diffuse reflection UV/Vis spectra from several graphitic carbon nitrides derived from melamine doped with barbituric acid. The relative mole fraction $\chi_{comonomer}$ of the used barbituric acid is given by g-CN$_{BA\_X}$*

Moreover, for these compounds UV/Vis examinations were done. In order to interpret the results straight forward, only the integrated areas of the UV/Vis spectra are considered in Figure 4-6 as a column diagram. These columns represent the overall reflected light δ. (The method was described in section 4.1.1, page 34.) A high value for the overall reflected light is caused by a material that absorbs light less efficient than a material with a low value (additional exemplary spectra are shown in the appendix in Figure 0-3). The trend of the spectra proves that with increasing amount of copolymerized barbituric acid the light absorption of the material increases. The overall reflected light δ of pure g-C$_3$N$_4$ is with 328 nm almost three times higher than 114 nm of the g-CN$_{BA\_8.2}$ (MA). This behavior was expected regarding the reported studies about copolymerization.[71] Speaking of semiconductor materials, a lower overall reflected light corresponds with a higher irradiation energy uptake. Light absorption increases with higher content of barbituric acid. Hence, the introduction of barbituric acid and a red shifting of the UV/Vis spectrum are sufficient.[70] This might result in an enlarged photocurrent and activity for photocatalytic water splitting.

Table 4-5: Elemental analysis regarding g-CN$_{comonomers\_x}$

Compound	χ_C / mol-%	χ_N / mol-%	χ_S / mol-%	C/N Ratio
g-CN$_{CA\_1}$ (MA)	40.2	59.8	-	0.67
g-CN$_{CA\_5}$ (MA)	40.2	59.8	-	0.67
g-CN$_{CA\_1}$ (DCA)	40.2	59.8	-	0.67
g-CN$_{CA\_5}$ (DCA)	40.1	59.9	-	0.67
g-CN$_{2MP\_1}$ (MA)	40.1	59.9	0.0	0.67
g-CN$_{2MP\_5}$ (MA)	41.4	58.6	0.0	0.71
g-CN$_{2MP\_1}$ (DCA)	40.4	59.6	0.0	0.68
g-CN$_{2MP\_5}$ (DCA)	42.8	56.8	**0.4**	0.75
g-CN$_{2TB\_1}$ (MA)	40.3	59.7	0.0	0.68
g-CN$_{2TB\_5}$ (MA)	42.4	57.4	**0.2**	0.74
g-CN$_{2TB\_1}$ (DCA)	40.5	59.5	0.0	0.68
g-CN$_{2TB\_5}$ (DCA)	42.6	57.1	**0.3**	0.75
g-CN$_{TTC\_1}$ (MA)	39.7	60.3	0.0	0.66
g-CN$_{TTC\_5}$ (MA)	39.9	60.1	0.0	0.66
g-CN$_{TTC\_1}$ (DCA)	39.8	60.3	0.0	0.66
g-CN$_{TTC\_5}$ (DCA)	39.8	60.2	0.0	0.66
g-CN$_{6MI\_1}$ (MA)	40.2	59.8	0.0	0.67
g-CN$_{6MI\_5}$ (MA)	41.6	58.2	**0.2**	0.72
g-CN$_{6MI\_1}$ (DCA)	40.2	59.8	0.0	0.67
g-CN$_{6MI\_5}$ (DCA)	42.9	56.7	**0.4**	0.76

Since it was proven that simple, straight forward modification can be carried out by copolymerizing barbituric acid, this technique should be transferred to other heteroatoms. It was shown previously that sulfur was introduced to the structure of g-C$_3$N$_4$ with much effort.[71] In order to introduce new dopants for the improvement of graphitic carbon nitrides easily, several copolymers and ratios of copolymers were explored. Investigations towards other copolymers were carried out. Therefore, compounds **7** to **11** (Figure 4-4) were copolymerized with the standard precursors. All derived carbon nitrides were examined with powder XRD, which showed typical signals for graphitic carbon nitride at 13° and 27°. The aim of these experiments was to prove the successful introduction of heteroatoms like carbon, sulfur and oxygen.

Elemental analysis results and UV/Vis interpretations are depicted in Table 4-5 and Figure 4-7. The results from the elemental analysis indicate that the triazine derivative copolymers cyanuric acid and trithiocyanuric acid (**7** and **10**) hardly change the carbon to nitrogen ratio (g-CN$_{CA}$ and g-CN$_{TTC}$). Although it was not expected for cyanuric, acid copolymerization with trithiocyanuric acid showed no introduction of sulfur. Still, since it became obvious by DTA that copolymers influence the polymerization process, a positive effect regarding catalytic activity might occurs due to an altered copolymerization process.

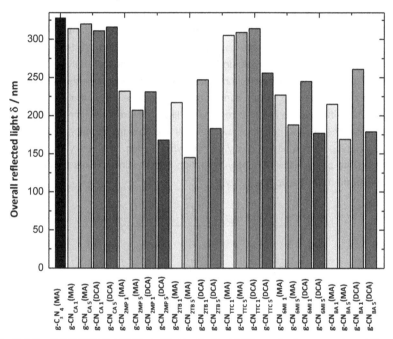

Figure 4-7: *Overall reflected light from diffuse reflection UV/Vis spectra from several graphitic carbon nitrides doped with various copolymers. Each value of a copolymerization process is depicted in a different color.*

To interpret the results from elemental analysis from copolymerization processes with compounds containing sulfur groups derived from pyridine (**8**, **9**, **11**) g-CN$_{2MP}$, g-CN$_{2TB}$ and g-CN$_{6MI}$ have to be pondered. Following the same consideration as for these compounds as for barbituric acid regarding the C/N ratio, about 60 % of the comonomers were introduced into the structure. Nevertheless, the sulfur mole fraction χ_S is far below that value, with just 0.0 % to 0.4 % sulfur. Although the introduction is not very efficient, it worked reasonably well with higher copolymer proportion. Furthermore, a better polymerization process can be seen with dicyandiamide than with melamine.

However, similar elucidation can also be obtained from integrated UV/Vis spectra, which symbolize the overall reflected light δ (Figure 4-7). A color code is used for each resulting carbon nitride. Compared to pure g-C_3N_4 all copolymerized variants have a higher light absorption. Furthermore, trends for each copolymer could be observed. Cyanuric acid (red) leads to almost no change.

Graphitic carbon nitride derived from the precursors and 2-mercaptopyrimidin (blue) has the highest light absorption for a mole fraction $\chi_{comonomer}$ of 5 relative mol-% of the copolymer and with DCA. It is not surprising that there can be similar results be seen for other copolymers like 2-thiobarbituric (green), trithiocyanuric acid (grey), 6-Mercaptoisocytosine (purple) and barbituric acid (maroon). Although melamine seems to be the better precursor for some copolymers, in order to increase the light absorption of the material, still dicyandiamide has proven by elemental analysis to enhance the ability to introduce the heteroatom sulfur. According to elemental analysis, just pyrimidine derivatives with higher carbon contents reduce the overall reflected light strongly. Thus, the spectra are also red-shifted like it was reported for barbituric acid. Nevertheless, these experiments show that it is possible to introduce new elements into the graphitic carbon nitride easily by copolymerization. It is highly recommended to examine the possibilities of copolymerization further, since it was proven to be a straight forward modification possibility.

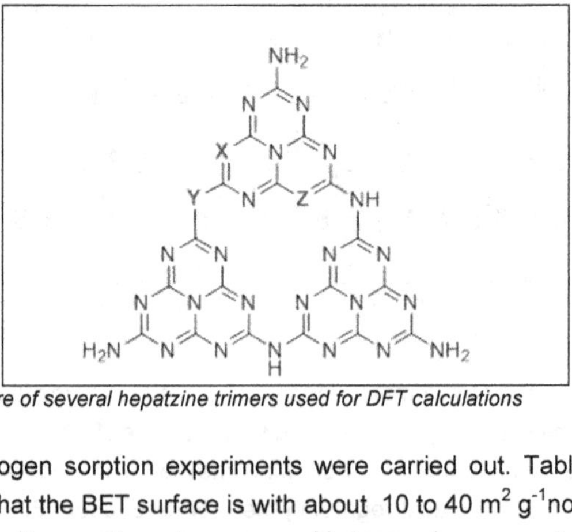

Figure 4-8: *Structure of several hepatzine trimers used for DFT calculations*

Furthermore, nitrogen sorption experiments were carried out. Table 0-1 (Appendix: page 82) shows that the BET surface is with about 10 to 40 $m^2 g^{-1}$ not small compared to other materials like zeolites, decreases with increasing amount of copolymerized material. Hence, the surface decreases, when dicyandiamide is used as precursor instead of melamine. Notably, the BET surface for both basic material increases when they are activated by heating them under reflux in hydrochloric acid. The results from the nitrogen sorption can be comparable with known BET surfaces.[49]

Density functional theory (DFT) calculations were performed to elucidate the electronic properties of heptazine rings doped with carbon, sulfur and oxygen. DFT is a powerful tool to predict HOMO and LUMO and a vast of other properties like molecule geometry. In this thesis, it was performed to give theoretical background to the doping experiments, which have been conducted.

The results of HOMO and LUMO energy calculations are displayed in Figure 4-9. The theoretic calculations were done according to literature.[83] The calculated gap between HOMO and LUMO cannot be evaluated quantitatively. On the one hand, even for small molecules, the DFT expectations are presumably incorrect due to the limitations given by assumption made by that method like setting the temperature to 0 °K and modeling in the gas phase. On the other hand, limitations are made by computing only one molecule instead of a solid bulk phase.

Table 4-6: Terms for DFT-calculated molecules

Term	X Position	Y Position	Z Position
DFT I	C	NH	N
DFT II	N	NH	O
DFT III	N	O	N
DFT IV	N	NH	S
DFT V	N	S	N
DFT VI	N	O	S
DFT VII	N	S	O
DFT VIII	N	NH	N

Trends and results can be discussed just qualitatively. However, these methods have proven to be suitable for such a question. Besides, by calculating the HOMO and LUMO of single molecules, a trend for the energies of the valence band and the conduction band might be predicted. Generally, a similar shift of the HOMO and LUMO can be observed depending on the molecular doping. For example, DFT I shows a slightly higher energetic HOMO while the energy of the LUMO is increased a little further compared to the pure heptazine (DFT VIII). Considering doping the semiconductor properties g-C_3N_4, this would correspond in a classical way to a negative doped semiconductor like silicon doping with phosphorous. Therefore, the results from UV/Vis spectroscopy seem reasonable. By an additional energy level which is located under the conduction band of the semiconductor, the effective band gap gets allegedly smaller. Thus, the UV/Vis spectra of this class of compounds will be red-shifted. Nevertheless, keeping classical semiconductor doping in mind, just a few units within the polymer are needed for an efficient doping. Because of the termination of the semi-conductor properties by adding to much dopant, the concentration of carbon doping within the catalyst has to be controlled properly.

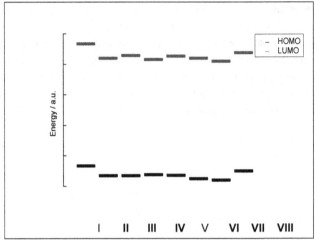

Figure 4-9: *Results of DFT computed – due to the nature of DFT calculations, just trends can be delineated*

Discussing DFT II to DFT VII, apparently all derivatives of heptazine show similar HOMO and LUMO properties; because of this, they will be discussed as a group. Within this group, the influence of heteroatoms such as sulfur and oxygen was calculated. Due to the absence of information of the incorporation of heteroatoms into the structure of graphitic carbon nitride, different positions for these atoms were assumed. Following the same arguments named DFT I and negative doping, these kind of doping of graphitic carbon nitride corresponds to positive doped semiconductors. This is contraindicated, because oxygen and sulfur are supposedly electron donors. Nevertheless, the conduction band, represented by the calculated LUMO, is lowered in energy. Doping the material with just a few monomers could be essential for maintaining the semiconductor properties. Unlike to carbon doped heptazine systems, at least the models DFT III and DFT V show a structure, which does not interrupt the mesomeric system of the heptazine ring. Since the corresponding structures have no substitution at the heptazine ring, altering the bridge position supposedly will not change the conformation of the compound. Thus, the original geometry of this class of compounds can be retained.

In order to investigate the influence of copolymerization on the catalytic activity of graphitic carbon nitrides towards light-driven hydrogen evolution, catalytic experiments were performed. The photocatalytic hydrogen evolution from water was investigated. The relative hydrogen evolution rates are displayed in Table 4-7. For clarity, just the relative hydrogen evolution rates were plotted. The results are normalized to g-C_3N_4 (MA), hence, 100 % are 140 μmol h^{-1}.

Table 4-7: Relative hydrogen evolution rate of copolymerized graphitic carbon nitrides

Compound	Relative hydrogen evolution rate / %
g-$CN_{BA\_0.5}$ (MA)	49.8
g-CN_{BA_1} (MA)	31.2
g-$CN_{BA\_2.1}$ (MA)	29.3
g-CN_{BA_5} (MA)	14.4
g-$CN_{BA\_8.2}$ (MA)	1.1
g-CN_{BA_1} (DCA)	29.2
g-CN_{BA_5} (DCA)	3.9
g-CN_{CA_1} (DCA)	150.9
g-CN_{CA_5} (DCA)	123.9
g-CN_{2MP_1} (DCA)	35.3
g-CN_{2MP_5} (DCA)	7.8
g-CN_{2TB_1} (DCA)	51.1
g-CN_{2TB_5} (DCA)	13.6
g-CN_{TTC_1} (DCA)	96.0
g-CN_{TTC_5} (DCA)	66.5
g-CN_{6MI_1} (DCA)	31.6
g-CN_{6MI_5} (DCA)	12.8

To examine the influence of carbon or barbituric acid doping, respectively, on graphitic carbon nitride derived from melamine, the hydrogen evolution rate of all synthesized semiconductor materials were tested. It is obvious that there is still a significant hydrogen evolution arising from graphitic carbon nitride derived from barbituric acid copolymerized melamine. Nevertheless, the catalytic activity of the material decreases with the introduction of carbon provided by the copolymer. The relative hydrogen evolution rate drops from 50 % to 1% with increasing amount of introduced carbon.

Regarding the water splitting results and the results from DFT calculation and elemental analysis, this might be caused by too high carbon doping. Investigating the loss of activity between g-CN_{BA_5} (MA) and g-$CN_{BA\_8.2}$ (MA), changing the ratio of additional carbon to heptazine from below to above 1.0, the amount of produced hydrogen drops drastically. From the collected data can be concluded that, in order to gain a more active catalyst the ratio of barbituric acid to melamine must be adjusted more finely between pure melamine and 0.5 relative percentages comonomers mixed with melamine.

Because of the results from elemental analysis and UV/Vis spectrometry, just the carbon nitrides from dicyandiamide were analyzed with respect to their catalytic activity. All copolymer-derived carbon nitrides show a different hydrogen evolution rate. Compared to the melamine based material g-CN$_{BA\_1}$ (DCA) and g-CN$_{BA\_5}$ (DCA) show similar results.

The copolymerization of cyanuric acid leads to a higher hydrogen evolution rate which is 1.5 times higher compared to the basic material g-C$_3$N$_4$. This might derive from oxygen introduction or from changing the polymerization process in general. From the latter could be concluded that the polymerization process progresses better because of a eutectic mixture.

However, this was the only compound, which led to an improved catalytic material. The semiconductors g-CN$_{2MP\_X}$ (DCA), g-CN$_{2TB\_X}$ (DCA) and g-CN$_{6MI\_X}$ (DCA) showed no enhanced catalytic activity. The photocatalytic activity of these materials toward photocatalytic hydrogen evolution from water decreases compared to g-C$_3$N$_4$: around 50 % for lower comonomer amount and around 10 % for the higher conmonomer amount

Because g-CN$_{2MP\_X}$ (DCA) and g-CN$_{6MI\_X}$ (DCA) had basically the same results in their performance compared to barbituric acid copolymerized dicyandiamide, this cannot be seen as an improvement. Anyhow, by copolymerizing thiobarbituric acid with dicyandiamide (g-CN$_{2TB\_1}$ (DCA)), the hydrogen evolution was enhanced from about 30 % to 50 %, when it is compared to g-CN$_{BA\_1}$ (DCA). This shows that even carbon doped graphitic carbon nitride can be improved by the introduction of sulfur heteroatoms. Double heteroatom doping is possible.

In this section, graphitic carbon nitrides from pure precursors differ from materials which were derived from a copolymerization process. Since the catalyst differs slightly depending on which precursors were used, it seemed to be logical that analyses towards polymerization processes of both needed to be done. The first Maldi-MS experiment ensured the presence of heptazine rings in g-C$_3$N$_4$. By DFT calculations and thermal analysis a possible explanation for sufficient heteroatom introduction and catalytic improvement could be given.

A new method was found to introduce sulfur atoms into the structure of the catalyst. As it was discussed above, the results for the copolymerization imply that it is highly effective and straight forward to introduce carbon and sulfur into the structure of graphitic carbon nitride. Therefore, other heteroatoms can be used to dope the semiconductor. Copolymerization promises higher effectiveness towards easy graphitic carbon nitride improvement.[69,70,72]

4.2 Sensitizing

4.2.1 Immobilization of Dyes

Grätzel came up with idea of sensitizing semiconductors via immobilization of dyes to enhance the quantum yield of photogenerated electrons.[18] Several dyes were immobilized on TiO_2 for DSSC in order to get a higher photocurrent.[74] Following these considerations, this strategy was used to enhance the activity of graphitic carbon nitride towards photocatalytic water splitting. In previous examinations adsorption and impregnation of dyes onto g-C3N4 was used to dye-sensitize the photocatalyst towards hydrogen evolution from water.[75] Naturally covalent bonding of dyes with graphitic carbon nitride could improve the stability and the activity of the catalyst.

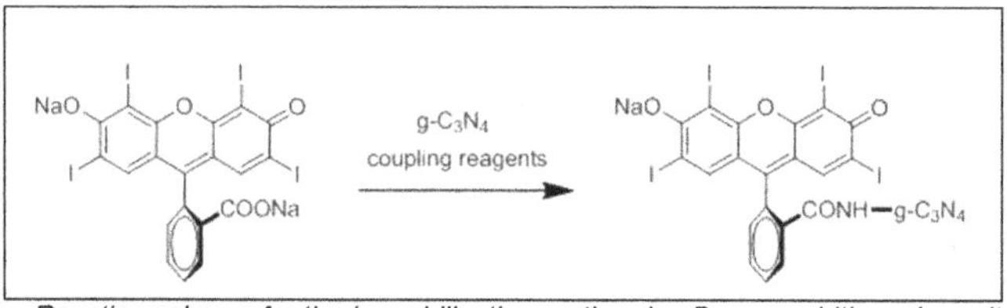

Figure 4-10: *Reaction scheme for the immobilization erythrosine B on graphitic carbon nitride*

Recently developed and presented in this thesis,[84] for properly modified g-C3N4, DIC and DMAP were used as coupling reagents in a highly diluted organic solution (Figure 4-10). This is possible, because g-C3N4 has free amine groups. Classically, this can be used to build an amide. Hence, the amount of erythrosine B immobilized and its stability on the catalytic material should be improved compared to electrostatic adsorption. Erythrosine B was used, because it has one of the highest known extinction coefficients with over 102000.
Still, another new method was explored for sufficient dye modification: electron beam treatment. This strategy was used in the past to immobilize small organic molecules on polymer surfaces like membranes or fibers. Therefore, this was supposedly a great tool for surface modification of graphitic carbon nitride which has aspects of an organic compound.[76]

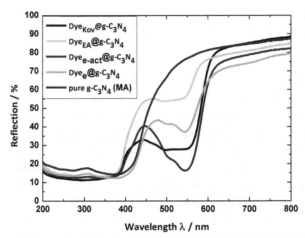

Figure 4-11:   *UV/Vis spectra from four dye modified catalysts derived from melamine and pure g-C₃N₄*

In Figure 4-11 the UV/Vis spectra from different erythrosine B modified graphitic carbon nitrides are shown. The black and grey curves show exemplary the difference between the two immobilization techniques. Erythrosine B absorbs light between 408 and 600 nm. In both variants, the peaks caused by erythrosine B are flatted. The reason for the flattening might be the acidic surface. Though, covalently bound dyes seem to be immobilized with much higher concentrations compared to the method using electrostatic adsorption. Because of the immobilization of the dye, the absorption spectrum might be altered. Regarding dye modification by means of electron beam treatment (reddish curves), it becomes obvious that an activated graphitic carbon nitride is able to concentrate the dye much more efficient on the surface. Because the same amount of erythrosine B was for all experiments equivalent, it was expected that the UV/Vis spectra of these compounds do not differ. Surface properties of activated graphitic carbon nitride enhance the proper immobilization of the dye, instead of a possible diffusion of the dye into the material or the unserviceable attachment on the solid surface. Figure 4-11 shows that the spectrum of graphitic carbon nitride is blueshifted during activation of the catalyst with hydrochloric acid.

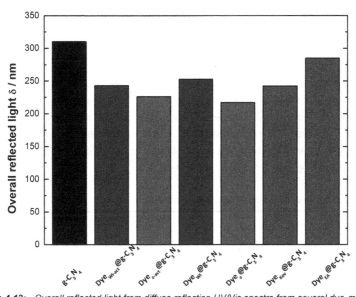

Figure 4-12:   *Overall reflected light from diffuse reflection UV/Vis spectra from several dye-modified g-C₃N₄*

Since the overall diffuse reflection is an important value to gain information about a semiconductor used for photocatalysis, the reflection of dye-sensitized graphitic carbon nitrides were integrated as it was discussed in section 4.1.1. The results can be seen in Figure 4-12 as a column diagram. Unlike to the procedure introduced in section 4.1.1 (page 34.), the total diffuse reflection was integrated from 300 nm instead of 200 nm to 800 nm. Thus a smaller overall reflected light is observed. This was done to take into account that sensitizing was done to shift the spectra into the visible range of the light. Since the black column represents the spectrum or pure graphitic carbon nitride, all sensitizing variants had an impact on the sum of the overall reflected light.

It can be assumed that the overall reflected light δ corresponds to the amount of surface bound dye. Accordingly, the materials have a higher light absorption of visible light which is for all material around 250 nm. Wetness impregnation method had higher dye content than the electrostatic adsorption and the covalent bonding.

Even though, for electron beam treatment and wetness impregnation 10 mg dye and 500 mg g-C₃N₄ were used, different amount of surface bound dye is immobilized by each of the techniques. Interestingly, the electron treatment has definitely a positive impact on surface modification with dyes. Furthermore, when it comes to analyze the success of covalent bonding versus electrostatic adsorption, clearly, Dye$_{Kov}$@g-C₃N₄ has a much higher amount of dye immobilized on the surface than its non-covalent counterpart. Hence, the covalent bound erythrosine B is much more stable compared to non-covalent bound dyes.

Originally, it was the aim to analyze the binding of the dye on the solid surface. Therefore, again Maldi mass spectrometry of $Dye_{Kov}@g\text{-}C_3N_4$ (MA) was done which is shown in the appendix (Figure 0-7, page 83). Unfortunately, the spectrum gives hardly information how the dye is bound to the surface. Still, fragments of graphitic carbon nitrides can be found as well as peaks which possible can be assigned to decomposition products of dye. The main signal at 393 m/z could derive from an erythrosine B fragment, which lost three iodine atoms and the carboxylic acid while the sodium ions are dissolved during the immobilization process.

Table 4-8: Hydrogen evolution rate of dye modified pure g-C₃N₄

Compound	Relative hydrogen evolution rate / %
$g\text{-}C_3N_4$ (MA)	100
$Dye_{EA}@g\text{-}C_3N_4$ (MA)	241
$Dye_{Kov}@g\text{-}C_3N_4$ (MA)	260
$Dye_{EA}@g\text{-}C_3N_4$ (DCA)	235
$Dye_{Kov}@g\text{-}C_3N_4$ (DCA)	255
$Dye_{e}@g\text{-}C_3N_4$ (MA)	630
$Dye_{e\text{-}act}@g\text{-}C_3N_4$ (MA)	115
$Dye_{Wl}@g\text{-}C_3N_4$ (MA)	506
$Dye_{Wl\text{-}act}@g\text{-}C_3N_4$ (MA)	249

The setup for the photocatalytic experiments was changed for the dye sensitized catalyst. Thus a glass filter was installed to improve the hydrogen evolution rate with visible light. This was done to use only the visible part of the irradiation spectrum from the Hg medium pressure lamp. Thus, the hydrogen evolution of pure graphitic carbon nitride was measured again under these conditions. Only 15.9 µmol H_2 were produced in one hour which is 11.4 % of the evolution rate under UV conditions. In order to have an initial overview, all results for the hydrogen evolution of the dye-sensitized $g\text{-}C_3N_4$ are normalized to that value and shown in Table 4-8. It becomes clear that dye-sensitizing is highly successful for enhancing the hydrogen production under visible light. Covalent and electrostatic adsorption modification almost triples the hydrogen production. In the average, additional 30 µmol evolved per hour. $g\text{-}C_3N_4$ with electrostatic adsorbed dye has a slightly higher hydrogen evolution rate compared to covalent bound erythrosine B. After one hour irradiation, covalent modified samples are unaltered, while the dye which was just adsorbed was probably photocatalytically decomposed. This behavior might be explained that dye, which is adsorbed on the surface, is highly active, but susceptible for photogenerated electrons for decomposition or reduction, respectively. On the opposite, covalent modified graphitic carbon nitride was not altered very well during the radiation of the samples. Loosely bound dye might be a better energy transmitter than tightly immobilized dye. Still, the advantage is given by the higher stability.

Regarding the modifications that were done with electron beam treatment, the results from hydrogen evolution experiments do not allow a simple analysis. Anyhow, the relative hydrogen evolution increased from 115 % to over 600 %, which is an enormous activity change.

The not activated catalysts perform better than the activated ones. Nevertheless, this effect is stronger after electron beam treatment and weaker after the modification of the catalyst just by wetness impregnation. It is assumed that treating a mixture of graphitic carbon nitride and erythrosine B with electron induces a stronger binding of the dye on pure g-C_3N_4. Hence, a decomposition of the active photocatalytic species occurs, when graphitic carbon nitride is heated in concentrated hydrochloric acid for the activating of the binding sites. Nevertheless, it seems that this new method is a good way for the immobilization of dyes on graphitic carbon nitride. Though stabilization tests were not conducted, detailed test promise a good chance for discovering the actual nature of dye-modification by electron beam treatment.

Overall, the results show that the photocatalytic hydrogen evolution from water was enhanced due to dye-sensitizing of g-C_3N_4. Covalent modification of the semiconductor leads to a higher catalytic activity regarding the higher stability of the immobilized dye. The hydrogen evolution rate doubled due to modification what is a huge step towards photocatalytic hydrogen evolution under visible light. Covalent modification and highly catalyst enhancement has not been reported for g-C_3N_4 with the minor amounts of dye used in the conducted experiments.[75,84]

4.2.2 Dependency on Defects

Because the covalent dye modification was successful, this kind of modification was attempted to be expanded on graphitic carbon nitride derived from melamine copolymerized with barbituric acid. The examination based in results shown preliminary.[84] Covalent bonding of dyes with graphitic carbon nitride was discussed in section 4.2.1. Thus, it was the aim to study, whether a covalent bonding of graphitic carbon nitrides which were derived from melamine and different amounts of barbituric acid would show a higher affinity for this modification. Of course, the focus was on understanding trends for future improvement of the catalyst.

This was a promising approach, because the introduction of additional carbon into the heptazine ring should provide different chemical environment. Thus, there might be two possibilities. On the one hand, steric hindrances – caused by an additional carbon atom compared to graphitic carbon nitride derived from pure melamine – could reveal possible adsorption sites for erythrosine B. On the other hand, the altered structure might lead to an enhanced number of possible bonding sites for the dye to be covalently bound to the solid surface. Therefore, again the diffuse reflection UV/Vis spectra were integrated and the overall reflected light δ was determined (Figure 4-13).

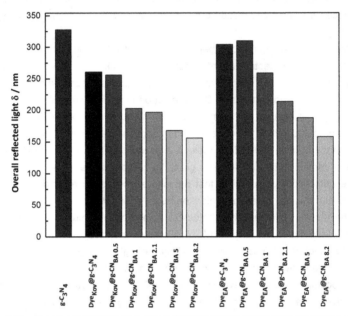

Figure 4-13:   *Overall reflected light from diffuse reflection UV/Vis spectra from several dye-modified g-C$_3$N$_4$ copolymerized with barbituric acid.*

All modifications have a higher overall light extinction than the basic material (Figure 4-13). Comparing covalent versus electrostatic dye immobilization, obviously, the dye content of covalent modified graphitic carbon nitrides preponderates over the amount of immobilized dye by the method of electrostatic adsorption. Nevertheless, there was no correlation found between the amount of copolymerized barbituric acid and the amount of immobilized erythrosine B.

Table 4-9: *Hydrogen evolution rate of dye modified g-C₃N₄ from melamine or dicyandiamide and barbituric acid*

Compound	Relative hydrogen evolution rate/ %
Dye$_{Kov}$@g-C$_3$N$_4$ (MA)	241
Dye$_{Kov}$@g-CN$_{BA\_0.5}$ (MA)	68
Dye$_{Kov}$@g-CN$_{BA\_1}$ (MA)	42
Dye$_{Kov}$@g-CN$_{BA\_2.1}$ (MA)	52
Dye$_{Kov}$@g-CN$_{BA\_5}$ (MA)	25
Dye$_{Kov}$@g-CN$_{BA\_8.2}$ (MA)	13
Dye$_{Kov}$@g-CN$_{BA\_1}$ (DCA)	56
Dye$_{Kov}$@g-CN$_{BA\_5}$ (DCA)	24

Dye modified g-CN$_{BA\_x}$ were tested towards their activity in photocatalytic water split-ting under visible light (Table 4-9). Indeed, the catalytic activity decreases strongly with the amount of copolymerized barbituric acid. 70 % relative hydrogen evolution for the best Dye modified g-CN$_{BA\_x}$ confirm an insufficient photocatalytic activity. Never-theless, the catalytic performance does not fall as much as shown for unmodified sys-tems from precursors and copolymers (section 4.1.2). These two effects are in con-tradiction to each other. While the dye transmits more energy to the semiconductor, the band structure of the semiconductor is disturbed by the copolymer as it was dis-cussed above. Because dye-sensitizing of graphitic carbon nitride is more efficient for copolymerized material, supposedly, the dye has to transmit energy depending on the amount of the copolymerized barbituric acid. However, changing the precursor to di-cyandiamide has hardly an influence.

The experiments showed the influence of the amount of barbituric acid used in the copolymerization on the affinity of graphitic carbon nitride being modified by dyes. This is a first hint that dye immobilization is depending on the structure of g-C$_3$N$_4$, which has not been reported yet. Furthermore, it was known that mesostructuring leads to a better impregnation of dyes.[75] Combining both approaches might be a powerful technique for dye-sensitizing of g-C$_3$N$_4$.

4.3   Nanoparticle Deposition

4.3.1   Deposition of Noble Metal Compounds

Since the importance of nanostructuring for photocatalysis and, beyond that, for pho-tocatalytic water splitting was reported and reviewed with a certain attention,[19] the interest grew to study the effects of nanoparticle deposition. The standard procedure for deposition of noble metals on graphitic carbon nitride is in-situ deposition by UV-irradiation of the semiconductor in noble metal salt solutions.[47] This procedure results in a platinum, rhodium and other noble metal modified photocatalyst. It was the aim, to establish other means for nanoparticle deposition, which do not need UV light for a sufficient modification. Supposedly, methods are useful, which provide a high controllability of shape and size of the deposited metals.

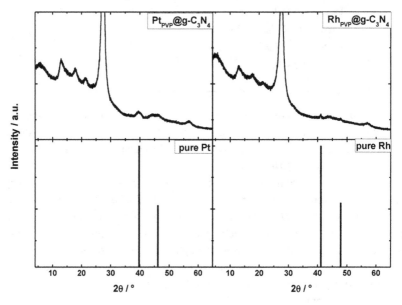

Figure 4-14: XRD patterns from NM$_{PVP}$@g-C$_3$N$_4$

Therefore, several experiments to deposit nanosized noble metal compounds on the solid surface of graphitic carbon nitride were carried out. Polyvinylpyrrolidone (PVP) mediated size-controlled synthesis of platinum and rhodium (NM$_{PVP}$@g-C$_3$N$_4$) modified graphitic carbon nitride was carried out.[33] A noble metal salt was reduced by ethylen glycol and stabilized in ethanol with PVP. PVP was burned off from the resulted nanoparticles, after the electrostatic adsorption of the particles on g-C$_3$N$_4$. It was the aim to establish a non-irradiation technique for this kind of catalyst. Furthermore, light deposition of these promoters was done (NM$_{light}$@g-C$_3$N$_4$), in order to compare the efficiency of the synthesis routes, this was performed under visible light. These elemental materials are known to promote hydrogen evolution on semiconductors. Furthermore, chromium and cobalt oxide were used to investigate the increase of hydrogen production by reducing the oxygen overpotential. All investigations were done with g-C_3N_4.

At first, the elemental noble metal deposition will be discussed. In Figure 4-14, the XRD patterns of the nanoparticle-modified graphitic carbon nitrides are shown, which were synthesized by means of PVP-mediated size control. As it is indicated by the displayed XRD intensities of pure platinum and rhodium, a PVP-mediated noble metal deposition was successful. Since it was the aim to develop a new strategy to modify g-C_3N_4 with noble metal compounds, comparative classical experiments were carried out.

For comparison, standard photocatalytic water splitting experiments with carbon nitrides are conducted with added noble metal salts like H_2PtCl_6 or $RhCl_3$. During the process of hydrogen evolution, probably by excited electrons from conduction band of the semiconductor g-C_3N_4, the Pt^{2+} and Rh^{3+} ions are reduced to their Pt^0 and Rh^0 metal clusters.

The results from noble metal deposition by visible light will be discussed later on. Nevertheless, the XRD data and TEM images for Pt and Rh (shown in the appendix - Figure 0-9 to Figure 0-11), implies a well dispersed precipitation of these noble metals by that method. The particles might have a diameter below 3 nm.

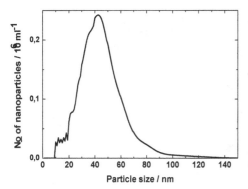

Compound	Maximum of the size distribution / nm
Pt_{PVP} (ethanol)	43
Rh_{PVP} (ethanol)	41
s-Co_3O_4 (water)	n.a.
m-Co_3O_4 (water)	n.a.
l-Co_3O_4 (water)	53

left **Figure 4-15:** *Exemplary results from nanoparticle tracking analysis for Pt_{PVP} nanoparticle*
right **Table 4-10:** *Tabulated results from nanoparticle tracking analysis*

The synthesis of NM_{PVP}@g-C_3N_4 was performed according to noble metal nanoparticle deposition on GaN and ZnO.[33] In the process of modification, nanoparticles are synthesized with a polyvinylpyrrolidone shell. Thus, the nanoparticles can be stabilized in glycol, ethanol or other polar solvents. The elucidation for a sufficient stable nanoparticle suspension in ethanol was done by nanoparticle tracking analysis. The results from the left Figure 4-15 and the right Table 4-10 prove that NM nanoparticles were sufficiently stabilized in ethanol. Although Rh_{PVP} and Pt_{PVP} have, unexpectedly, a relative high average diameter of about 40 nm, the polymeric PVP shell is supposed to have the utmost contribution for that.

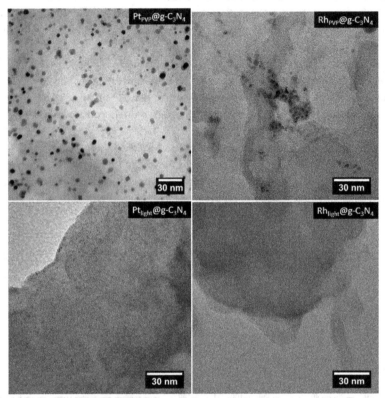

Figure 4-16:   *TEM images from platinum and rhodium modified graphitic carbon nitride. The upper left figure shows spherical platinum NPs and the upper right shows agglomerations of rhodium NPs; both by PVP mediated NP deposition onto g-C₃N₄ (NM_{PVP}@g-C₃N₄). The lower images show graphitic carbon nitrides which were modified by light deposition of platinum (left) and rhodium (right) (NM_{light}@g-C₃N₄). The noble metals are hardly detectable.*

In order to clarify, whether the electrostatic adsorption of noble metal PVP nanoparticle and the subsequent burning off of the polymer was successful, transmission electron microscopy (TEM) experiments were executed. The results depicted in Figure 4-16 prove that platinum and rhodium were sufficiently deposited by PVP mediated synthesis. Still, the platinum variant led to well dispersed, spherical platinum nanoparticle with a regularly particle diameter. Obviously, it can be assumed that this kind of deposition is far more suitable for platinum than rhodium on graphitic carbon nitride.

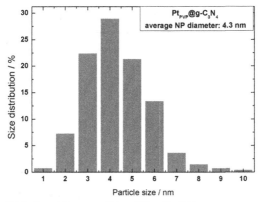

Figure 4-17: Size distribution of Pt nanoparticles of Pt_{PVP}@g-C_3N_4

Regarding the TEM images for PVP mediated deposition of platinum and rhodium on the solid surface of graphitic carbon nitride platinum particles were formed much more uniformly. The particles were evenly distributed compared to rhodium nanoparticles. TEM images reveal that both metals can be used for that kind of deposition. A size distribution of Pt particles was made due to the quality of the Pt_{PVP}@g-C_3N_4 TEM images suffices (Figure 4-17). Almost 30 % of the platinum nanoparticles have size range of 3.5 to 4.5 nm. Hence, the average size is 4.3 nm. Compared to nanoparticle UV-light deposition described in literature,[85] the particles size is about two times larger. Unfortunately, this advantage cannot be applied to Rh PVP mediated nanoparticle deposition. The size of the nanoparticles could hardly be determined by the resulted TEM images (Figure 0-8). Nevertheless, they seem to have minor size compared to Pt particles (estimated average particle size: about 2 nm). Even though, the advantage of the PVP mediated nanoparticle synthesis is the simple size-control of platinum and rhodium by changing the pH value during the synthesis of PVP coated noble metal NP.

Comparing the latter method with light deposition of nanoparticles, the advantage of the PVP mediated nanoparticle synthesis becomes more evident. The TEM images show that platinum hardly and rhodium not measurable get deposited in a sufficient way on the solid surface of g-C_3N_4. Thus, it would not be efficiently possible to deposit noble metal particles, if irradiation below 300 nm could not be applied to the system.

Therefore, PVP mediated synthesis of nanoparticle modified graphitic carbon nitride is supposedly a good variant to upscale to production of a useful semiconductor for hydrogen evolution from water splitting. The method provides spherical nanoparticles with defined size distribution. The size and the amount of deposited particles can be well regulated.

Table 4-11: *Elemental analysis by ICP-OES of noble metal modified g-C$_3$N$_4$*

Compound	ω_{Pt} / wt.-%	ω_{Rh} / wt.-%	ω_{Cr} / wt.-%
Pt$_{light}$@g-C$_3$N$_4$	2.2	-	-
Pt$_{PVP}$@g-C$_3$N$_4$	0.46	-	-
Rh$_{light}$@g-C$_3$N$_4$	-	0.84	-
Rh$_{PVP}$@g-C$_3$N$_4$	-	0.31	-
Cr$_2$O$_3$@g-C$_3$N$_4$	-	-	0.003
Cr$_2$O$_3$@Pt$_{PVP}$@g-C$_3$N$_4$	0.13	-	0.012
Cr$_2$O$_3$@Rh$_{PVP}$@g-C$_3$N$_4$	-	0.21	0.015

Regarding the elemental analysis by ICP-OES for all noble metal modified semiconductors (Table 4-11), it is noticeable that the mass fraction ω_{NM} of deposited noble metal nanoparticles is in the range of noble metals during in-situ photoreduction of about 0.05 wt.-%. The mass fraction ω_{Pt} of 2.2 wt.-% for Pt$_{light}$@g-C$_3$N$_4$ might be an artifact of the measurement. PVP mediated deposition gives just a medium noble metal mass fraction ω_{Pt} and ω_{Rh} of about 0.4 wt.-%. This is not in contradictory to the general aim enhancing the photocatalytic activity. Since the right size of the NPs can be crucial for the efficiency of photocatalytic water splitting, the influence of the promoter ω_{NM} towards hydrogen evolution will have to be discussed in the future.

In order to modify the nanoparticle further and enhance the photocatalytic activity of graphitic carbon nitride, it was tried to achieve a core-shell modification of the noble metal NPs with Cr$_2$O$_3$. Therefore, a photodeposition technique was applied using K$_2$CrO$_4$. This material should be photoreduced to chromium oxide and create a shell around rhodium and platinum particles. Regrettably, neither TEM images from the prepared samples nor XRD patterns revealed whether Cr$_2$O$_3$ was formed during that process or not. The results are shown in the appendix (Figure 0-9 to Figure 0-14). However, elemental analyses of chromium modified samples (Table 4-11) imply that the element was deposited on graphitic carbon nitride somehow. The mass fraction of chromium ω_{Cr} is about 0.01 wt.-% low. The chromium content rises when the deposition is tested on Pt and Rh modified samples by the order of one magnitude. Thus, there are two explanations. One the one hand, Cr$_2$O$_3$ was sufficiently synthesized, but it is not detectable by the used techniques. On the other hand, K_2CrO_4 was finely dispersed on the solid surface and not reduced, for that, it was not detected by XRD or TEM. Still, the amount of chromium is very low compared to the other noble metals. Interestingly, the amount of Rh and Pt decreases after the chromium modification attempts. Following a general understanding, noble metal NPs might be separated from the solid surface during modification.

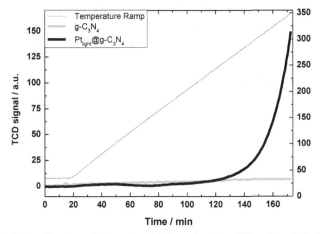

Figure 4-18: TPR profile of pure (light grey) and platinum doped graphitic carbon nitride (dark grey)

Although a sufficient deposition of platinum was proven, analyzing the stability and the properties under hydrogen treatment is most important for the catalytic reaction. Hence, temperature programmed reduction was carried out, which is depicted in Figure 4-18. Pure graphitic carbon nitride was compared with graphitic carbon nitride with light deposited platinum. Mild conditions were chosen, because of the possibly harming degassing process of the material at higher temperatures. Hence, TPR can be evaluated only qualitatively. Nevertheless, the noble metal shows huge influence towards the reducibility of g-C_3N_4 under a H_2 gas stream.

Pure graphitic carbon nitride shows no significant hydrogen turnover and therefore no reduction of the analyzed material within the displayed temperature area. In additional experiments, it was shown that the pure g-C_3N_4 had a positive TCD signal at temperatures above 500 °C. It was not investigated whether this is an effect from the decomposition or from the reduction of the material. While no sufficient hydrogen turnover occurred on the pure compound, platinum doped graphitic carbon nitride was reduced under comparatively mild conditions. The reduction of the material begins at 120 °C, below the decomposition temperature of the investigated material. Although it was not thermally treated further, the compound supposedly gets entirely reduced by additional heating. Thus, it is elucidated that platinum sensitizes graphitic carbon nitride for hydrogen reduction. Because the material is disintegrated during that process, a hydrogen spillover from platinum to g-C₃N₄ is plausible.

Hydrogen spill over is not very likely in an aqueous solution at low temperatures like 20 °C. More information about the electron transfer from H_2 over Pt to g-C₃N₄ than the imaginable decomposition process due to photocatalytic reduction of the semiconductor can be gained. It is likely that photogenerated electrons can be transferred to a proton in aqueous solution, due to lowered activation energy. This was proven by the TPR experiment.

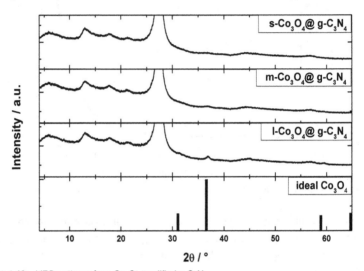

Figure 4-19: *XRD patterns from Co_3O_4 modified g-C₃N₄*

In the noble metal deposition experiments described so far, it was the aim to decrease the overpotential of hydrogen towards proton reduction. Yet there have been other developments for the photocatalytic hydrogen evolution from water. Cobalt (II) oxides and phosphates were introduces to increase the photocurrent of semiconductors and were discussed to reduce the overpotential for oxygen oxidation.[67] Herein presented is an investigation of the photocatalytic hydrogen evolution depending on the size of cobalt oxide nanoparticles. Cobalt oxide nanoparticles have previously demonstrated to increase the hydrogen evolution rate.[86]

Figure 4-19 shows the XRD patterns from Co_3O_4 which were electrostatically adsorbed onto the surface. Unfortunately, the integration of the cobalt oxide into the structure of graphitic carbon nitride by copolymerization was not successful. In both experiments, the batch with an aimed mass fraction ω_{Co3O4} of 1.0 wt.-% and with 0.1 wt.-% Co_3O_4, melamine burned off completely during the calcinations process and about 10 mg and 1 mg Co_3O_4 were retrieved. Nevertheless, an effective nanoparticle adsorption could be achieved by post-synthetic modification. Nanoparticle tracking analysis could not provide particle sized for all Co_3O_4 particles except for the attempt of the large ones (right Table 4-10).

Though, the results imply a larger particle size than expected,[77] a lack of stabilizability of the NP might product agglomeration. Cobalt oxide NPs were size-controlled synthesized by applying different amount of water and ethanol during the autoclave reaction. However, the XRD patterns prove the presence of cobalt oxide on g-C_3N_4. Therefore, it is possible to deduce the size of Co_3O_4 by using the scherrer equation, which is given in equation 4-1. For that XRD patterns were investigated thoroughly and signals were fitted to evaluate the half-width of a diffraction reflex β (Figure 4-19). By that method the particle size D_P was determined by the used wavelength λ, the line broadening at half the maximum intensity β and the Bragg angle of the particular peak θ_0. Thereby, the particle size of the nanoobjects was estimated.

$$D_p = \frac{0.94 \cdot \lambda}{\beta_{1/2} \cdot \theta_0} \tag{4-1}$$

Table 4-12: Characterization of the Co_3O_4 nanoparticles on g-C_3N_4

Compound	Nanoparticle size $_{Scherrer}$ / nm	Co wt.-% $_{ICP-OES}$
s-Co_3O_4@ g-C_3N_4	6	0.74
m-Co_3O_4@ g-C_3N_4	8	0.84
l-Co_3O_4@ g-C_3N_4	15	1.12

The calculated values and the results for elemental analysis by ICP-OES are shown in Table 4-12. It was the aim to synthesize nanoparticles with an adjustable size by using different ethanol/water ratios during the high-pressure autoclave reaction. By XRD measurements of Co_3O_4-modified g-C_3N_4, it was found out that the small particles (s-Co_3O_4) with a diameter of 6 nm are about half as small as the large particles (l-Co_3O_4) with a diameter of 15 nm. The diameter (8 nm) of the medium particles (m-Co_3O_4) is closer to the size of small particles. Because 1.00 wt.-% Co_3O_4 was applied, 0.73 wt.-% Co was the maximal applied mass fraction ω_{Co} on the catalytic material. As the elemental analysis validates, all of the nanoparticles were applied on g-C_3N_4. However, the percentage of cobalt exceeds the observed value.

Table 4-13: Hydrogen evolution rate of noble metal modified g-C$_3$N$_4$

Compound	Relative hydrogen evolution rate / %
Pt$_{light}$@g-C$_3$N$_4$	26.9
Pt$_{PVP}$@g-C$_3$N$_4$	90.2
Rh$_{light}$@g-C$_3$N$_4$	55.9
Rh$_{PVP}$@g-C$_3$N$_4$	60.3
Cr$_2$O$_3$@g-C$_3$N$_4$	76.0
Cr$_2$O$_3$@Pt$_{PVP}$@g-C$_3$N$_4$	26.9
Cr$_2$O$_3$@Rh$_{PVP}$@g-C$_3$N$_4$	6.8

The hydrogen evolution rate was measured for all noble metal modified graphitic carbon nitrides. The results are shown Table 4-13. During the analysis of the experimental data it must be kept in mind that graphitic carbon nitride without any noble metal modification is able just to produce about 3 μmol/h hydrogen. Consequently, only 2 % of the amount of hydrogen were produced, which evolves normally during irradiation g-C$_3$N$_4$ in noble metal salt solutions. Therefore, the standard method is in-situ modification with Pt salts giving a Pt^0 mass fractio ω$_{Pt}$ of 0.5 wt.-%. Herein, the latter was used as the standard reaction condition for the testing of the catalytic activity of all graphitic carbon nitrides, which were not modified with platinum or rhodium. Consequently, all noble metal depositions are suitable for the modification of graphitic carbon nitride with promoters towards hydrogen evolution. However, the catalytic activity differs with the used modification method. Although PVP mediated synthesis of Pt and Rh modified g-C$_3$N$_4$ reach not the catalytic activity of the standard modification with platinum, Pt$_{PVP}$@g-C$_3$N$_4$ has a relative hydrogen evolution of over 90 %. Compared to the visible light deposited Pt variant, this is an improvement of a threefold hydrogen evolution rate. Regarding the results for rhodium modified semiconductor, catalytic experiments reveal about the same effects for the lighter noble metal. Nevertheless, a reduced activity of Rh noble metal modified samples, which were prepared by the standard in-situ deposition showed a lowered activity as well. Therefore, the results are consistent. The positive effect of noble metals like rhodium or platinum can be explained by a reduction of the overpotential of the reaction from H$^+$ to H$_2$. Therefore, the photogenerated electrons from the conduction band of the semiconductor must be able to be transferred from g-C$_3$N$_4$ to the promoter.

In order to analyze whether the desired deposition of Cr_2O_3 had an influence on the catalytic activity of the basic material, the chromium modified samples were tested (Table 4-14). Cr_2O_3@g-C_3N_4 was tested with added Pt salt and Cr_2O_3@NM@g-C_3N_4 without. Nonetheless, it can be stated that chromium oxide somehow poisons the catalyst, since every chromium oxide-modified sample showed less activity than without modification. Because the method was only used to modify noble metal compounds on GaN based materials so far, the increase of the hydrogen evolution rate under irradiation which was found in earlier experiments,[33] might be caused by Cr doping of GaN or similar compounds.

Compared to previous results,[47] the presented experiments show sufficient deposition of nano-sized noble metal promoters without UV light for photocatalytic hydrogen evolution from water. This diminishes the drawback of in-situ NP deposition. The conducted experiments might help to develop an upscale for photocatalytic active g-C_3N_4.

Table 4-14: Hydrogen evolution rate of Co_3O_4@g-C_3N_4

Compound	Relative hydrogen evolution rate / %
s-Co_3O_4@g-C_3N_4	141
m-Co_3O_4@g-C_3N_4	67.4
l-Co_3O_4@g-C_3N_4	182

At last, Co_3O_4@ g-C_3N_4 samples were tested (Table 4-14). It was shown previously that Co^{II} salts or oxides have an impact on the photocatalytic activity of semiconductors.[86] This behavior might be explained by a reduced overpotential for oxidation reactions or an improved electron-hole separation followed by hole transmission from the valence band to the cobalt species. The Co_3O_4@ g-C_3N_4 samples were tested with Pt salts added to the reaction solution.

According to previous results, Co_3O_4 increases the hydrogen evolution rate of graphitic carbon nitride. Thus, it was the aim of the presented thesis to investigate the size-dependency of cobalt oxide nano particles, which were deposited on the solid surface of g-C_3N_4. As the results truly imply, such a dependency can be found. m-Co_3O_4@g-C_3N_4 and l-Co_3O_4@ g-C_3N_4 have a relative hydrogen evolution rate far beyond 100 % compared to g-C_3N_4. m-Co_3O_4@g-C_3N_4 shows just minor photocatalytic activity with 67 % of the basic hydrogen evolution. As it was already reported for several nanoparticle catalyzed reactions,[86] the size-activity relationship can undergo a minimum. Although cobalt oxide accelerates the reaction, this effect is stronger for the small and large nanoparticle with a size of 6 nm and 15 nm, respectively. Generally, such an effect is explained by the alternated percentage of different surface atoms. Still, the results might be an artifact from cobalt oxide nanoparticle giving different shapes caused by the autoclave reaction itself. A reduced activity for m-Co_3O_4@ g-C_3N_4 might derive from interruption of the platinum promoter. This means that the ratio of cobalt oxide and platinum or rhodium, respectively, must be tuned properly to enhance the photocatalytic hydrogen evolution from water. The conducted results for cobalt oxides follow previous experiments and enlarge the understanding of Co (II) promoters.[49] The size of promoting NPs influence considerably the photocatalytic activity towards hydrogen evolution. This fact might be transferred to other known promoters like Pt or Rh.[85]

4.3.2 Combination of Dyes Immobilization and Nanoparticles Deposition

Covalent dye immobilization was successful and the deposition of nanoparticles sufficed properly. To combine advantages of both strategies, pyrrole was oxidized by H_2PtCl_6 to polypyrrole. Regarding the reaction scheme (Figure 4-20), an anion can be introduced to the structure by non-covalent interaction with the positively charged polymer.

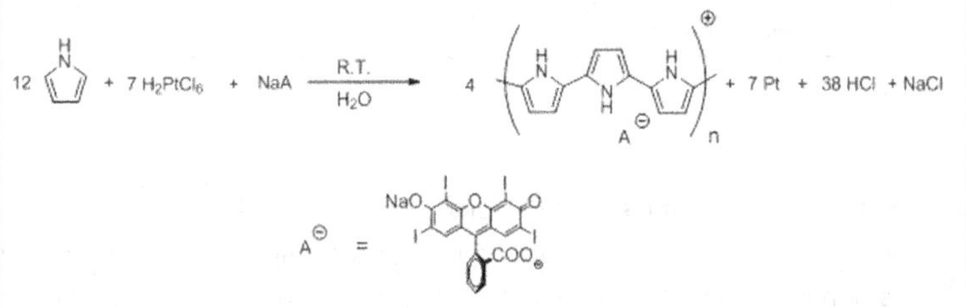

Figure 4-20: *Synthesis route for polymerization of Ppy onto g-C_3N_4*

Hence, erythrosine B might be able to bind to the polymeric chain of PPy. The ability of polypyrrole to transport electrons was used for dye-sensitized solar cells. In the context of photocatalytic water splitting this could lead to an increased activity by higher water conversion or decreased activity by accelerated electron-hole pair recombination. Furthermore, the deposition of platinum particles should occur. Hence, two modification steps should be combined to one.

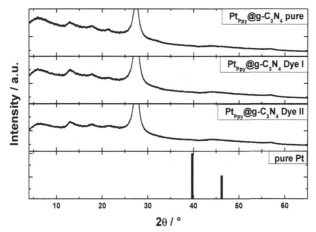

Figure 4-21: XRD patterns of platinum modified g-C₃N₄ by pyrrole modification process

Considering XRD patterns in Figure 4-21, it is obvious that no detectable elemental platinum was synthesized by the used method. There are two possible explanations. The modification and Pt^0 deposition was not successful or the elemental platinum is finely dispersed. Since all of the material was blackened during the process, it can be concluded that polypyrrole was deposited on the surface of g-C₃N₄ properly. Since no other oxidizing agent was applied, the Pt must be finely dispersed. Because elemental analysis (Figure 4-20) shows that almost all of the 6 % applied Pt was deposited on the graphitic carbon nitride, ICP-OES analysis might be a hint that maybe $PtCl_6^{2-}$ was introduced into the positively charged polypyrrole. Thus, it still should be able to be photoreduced during hydrogen evolution under irradiation. However, no significant hydrogen evolution was detected.

Therefore, the modification with polypyrrole must be regarded as a failure, but the proof of principle was shown. Polypyrolle modification is possible proven by elemental analysis and the reflection properties. Though, no influence of the amount of applied dye was observed, because the material was totally blackened during the polymeriza-tion process. Furthermore, polymerizing polypyrrole onto the surface of graphitic car-bon nitride led to a stabilized suspension, which has not been observed yet. For gra-phitic carbon nitride, which was modified with Ppy, neither mixing nor ultra-sonic treatment was required to have a stabilized solution in water.

Table 4-15: Elemental analysis by ICP-OES and hydrogen evolution rate

Compound	ω_{Pt} / wt.-%	Relative hydrogen evolution rate / %
$Pt_{Ppy}@g\text{-}C_3N_4$ pure	5.27	1.7
$Pt_{Ppy}@g\text{-}C_3N_4$ Dye I	4.86	1.9
$Pt_{Ppy}@g\text{-}C_3N_4$ Dye II	5.32	2.0

5   Conclusion and Outlook

In the presented thesis several modification of the semiconductor graphitic carbon nitride were realized. The aim of this thesis was the analysis of the activity of modified photocatalyst towards photocatalytic hydrogen evolution from water. Different characterization techniques were used to clarify the nature of the resulting changes. In order to modify the basic material, several synthesis routes were chosen to establish a sufficient in-situ and post-synthesis functionalization. Thus, three different approaches were examined.

At first, the impact on the polymerization process by dicyandiamide and melamine as precursors were studied. Therefore the formation of g-C_3N_4 was investigated. Though, graphitic carbon nitride showed hardly differences in XRD patterns or diffuse reflection UV/Vis spectroscopy, whether it was derived from melamine or dicyandiamide. Maldi mass spectrometry revealed that the material most possibly consists of polymers from hepatzine rings, as it was expected from previous results. The actual structure of the semiconductor is difficult to elucidate, but the conducted experiment enlarge the overall knowledge of this material.

However, the catalytic activity for the photocatalytic hydrogen evolution from water was slightly higher for dicyandiamide derived g-C_3N_4. The results follow the reported behavior of g-C_3N_4.

Furthermore, copolymers were used for a possible introduction of carbon, sulfur or oxygen into the structure of pure graphitic carbon nitride. To understand the copolymerization process, DTA, TG and DFT calculations were carried out. The enhancement of the photocatalytic activity might be an effect from n-doping the catalyst or by a better polymerization process due to a reduced melting point caused by a eutectic mixture. These experiments will help to understand how the reported red-shifting of the UV/Vis spectra occurs.

Applied comonomers were structurally similar to melamine and copolymerized with the precursor. For carbon doping barbituric acid was used. Melamine derivatives have heteroatom groups like thiols or alcohols instead of a primary amine. Accordingly, the derivatives were used to improve the photocatalytic activity of g-C_3N_4. XRD and UV/Vis spectroscopy validated the successful copolymerization of precursors with comonomers. Successfully, the sulfur content χ_S of graphitic carbon nitride was raised up to 0.4 mol-% for certain comonomers like thiobarbituric acid. Thus, a straight forward method for sulfur-doping was developed in the presented thesis.

For the material derived from dicyandiamide copolymerized with cyanuric acid the hydrogen evolution rate was enhanced 1.5 times of pure graphitic carbon nitride. This proves that copolymerization can enhance the photocatalytic activity even without carbon doping.

Secondly, graphitic carbon nitride was post-functionalized with erythrosine B in order to shift the absorption spectrum into the visible range of the light resulting in a higher quantum efficiency. Covalent bonding, electrostatic adsorption and electron beam treatment were used for dye immobilization.

In order to explore covalent dye bonding, diffuse reflection UV/Vis spectrometry was carried out. It was found that covalently bound erythrosine B is much more located on the surface compared to dye immobilization by electrostatic adsorption.

For these photocatalytic hydrogen evolution experiments, only visible light above 300 nm was used. It was found that graphitic carbon nitride is slightly more active if it is bound covalently to erythrosine B or just adsorbed on the surface. Nevertheless, the covalently bound dye exhibited much higher stability than its non-covalent counterpart. Thus, increasing the reaction time would reveal a higher light conversion to hydrogen by covalent bound dye. Dye-sensitized g-C_3N_4 by covalent immobilization has been proven to outperform previously reported dye impregnation in their photocatalytic hydrogen evolution from water under visible light.

Furthermore, this approach was used to study the influence of copolymerized barbituric acid on the improvement of graphitic carbon nitride by covalent dye immobilization. With a higher amount of copolymerized barbituric acid dye immobilization is more effective as for pure graphitic carbon nitride.

At last, the influence of different nano-sized noble metal particles, which were reported to act as a promoter for photocatalysis, was examined. Platinum and rhodium were supposed to be deposited on the surface of g-C_3N_4 to overcome the high overpotentials for hydrogen evolution from water. To compare different techniques, a novel PVP mediated nanoparticle deposition method and visible light deposition were examined. The known standard procedure involves just UV light deposition. A sufficient photocatalytic hydrogen evolution from water without Pt or Rh was not possible.

The PVP mediated synthesis provided an even size distribution for Pt on g-C_3N_4 with a maximum of 4.3 nm, which was investigated with TEM and XRD experiments. For platinum deposited on graphitic carbon nitride well shaped NPs were obtained. Even Rh modification was successful.

Compared to visible light deposition methods, the technique gave a higher photocatalytic activity towards hydrogen evolution from water. Thus, it is possible to deposit noble metal in the absence of UV light. PVP mediated synthesis provides a higher controllability of promoter composition and nanoparticle size than the reported methods.

Furthermore the deposition of Co_3O_4 was investigated towards photocatalytic water splitting. Electrostatic adsorption was used to modify g-C_3N_4. It was reported that Co (II) compounds can enhance the photocatalytic activity of semiconductors. Nanoparticle size investigations have not been reported yet.

In order to study the success of the nanoparticle deposition and the size of the cobalt oxide particles, they modified graphitic carbon nitride samples were examined by XRD.

For photocatalytic hydrogen evolution from water of cobalt modified g-C_3N_4, a size dependency was observed. The hydrogen evolution rate catalytic activity goes through a minimum at a medium particle size. Co_3O_4 was discussed to decreased the overpotential of the water oxidation reaction and stabilize electron holes leading to a higher photocurrent.

Regarding prospective researches, graphitic carbon nitrides are still a highly promising material for photocatalytic hydrogen evolution from water. The results presented in this work suggest that structure-activity dependency has to be further examined. Therefore, different synthesis strategies have to be compared.

Nonetheless, n-doping or p-doping of semiconductor will stay a good opportunity to alter the semiconductor properties towards a higher quantum efficiency. Thus, the concentration of copolymers must be adjusted very finely in order to get positive results for graphitic carbon nitride doping. With the described method other heteroatoms can be easily introduced into the structure of graphitic carbon nitride.

The principle of post-synthetic functionalization by covalent bonding was shown. Free amines of the graphitic carbon nitride can be used to perform classical organic chemistry reactions. Hence, bathochromic and auxochromic groups can be attached by post-functionalizing methods: for example the oxidation of $-NH_2$ to $-NO_2$ or an electrophilic aromatic substitution to introduce iodine or bromine.

Of course the deposition of promoters like Pt and Rh and other noble metals has to be further investigated. Due to the fact that the deposition method applying polyvinylpyrrolidone is able to produce nanoparticle with defined diameter, it is possible to check the influence of this parameter on the photocatalytic enhancement by size-determined noble metal promoters. Since the chromium oxide core-shell modification was not proven to be successful, XPS (X-ray photoelectron spectroscopy) could be an adequate method to determine the oxidation state of Cr.

At last, g-C_3N_4 could be a future option – modified with platinum – as a photocathode while this electrode is coupled to a photoanode like Fe_2O_3. Applying this idea for a system in which both compounds are suspended as nanoparticles and small molecules transferring the charge between both semiconductors, this would be the first suspended Z-scheme photocatalyst. Hence, a larger amount of the visible spectrum would be used for photocatalysis.

References

[1] a) K. Ledjeff, *Neue Wasserstofftechnologie*, C.F.Müller, Karlsruhe, 1986. b) U. R. von Eckart Lohse, "Eine neue Energiequelle", can be found under http://www.faz.net/aktuell/wissen/neue-energiequelle-politiker-traeumen-vom-wasserstoff-12545218.html, 2013.

[2] N. S. Lewis, D. G. Nocera, *Proceedings of the National Academy of Sciences* 2006, *103*, 15729–15735.

[3] a. H. H. Kreuter W, *Int. J. Hydrogen Energy*, *23*, 661–666.

[4] Ullmann, *Ullmann's Encyclopedia of Industrial Chemistry*, Wiley-VCH Verlag GmbH & Co. KGaA, Weinheim, Germany, 2000.

[5] B. Christian Enger, R. Løer deng, A. Holmen, *Applied Catalysis A: General* 2008, *346*, 1–27.

[6] J. M. Ogden, *Annu. Rev. Energy. Environ.* 1999, *24*, 227–279.

[7] A. Kudo, Y. Miseki, *Chem. Soc. Rev.* 2008, *38*, 253.

[8] M. Szklarczyk, *J. Electrochem. Soc.* 1990, *137*, 452.

[9] C. H. Hamann, W. Vielstich, *Elektrochemie*, Wiley-VCH, Weinheim ; Chichester, 1998.

[10] M. de Chialvo, A. Chialvo, *Journal of Electroanalytical Chemistry* 1994, *372*, 209–223.

[11] a) A. G. Chesbeck, *Nachr. Chem.* 2000, *48*, 173–174. b) N. Serpone, A. V. Emeline, *International Journal of Photoenergy* 2002, *4*, 91–131.

[12] M. G. Walter, E. L. Warren, J. R. McKone, S. W. Boettcher, Q. Mi, E. A. Santori, N. S. Lewis, *Chem. Rev.* 2010, *110*, 6446–6473.

[13] J. Ding, Y. Li, H. Hu, L. Bai, S. Zhang, N. Yuan, *Nanoscale Res Lett* 2013, *8*, 9.

[14] A. FUJISHIMA, K. HONDA, *Nature* 1972, *238*, 37–38.

[15] M. Fujihira, Y. Satoh, T. Osa, *Nature* 1981, *293*, 206–208.

[16] Y. Matsushita, T. Ichimura, N. Ohba, S. Kumada, K. Sakeda, T. Suzuki, H. Tanibata, T. Murata, *Pure Appl. Chem.* 2007, *79*, 1959–1968.

[17] F. Andrew Frame, E. C. Carroll, D. S. Larsen, M. Sarahan, N. D. Browning, F. E. Osterloh, *Chem. Commun.* 2008, 2206.

[18] B. O'Regan, M. Grätzel, *Nature* 1991, *353*, 737–740.

[19] F. E. Osterloh, *Chem. Soc. Rev.* 2013, *42*, 2294.

[20] a) Lewis, N. S.Nature 2001, 589, 365. b) J. R. Bolton, S. J. Strickler, J. S. Connolly, Nature 1985, 316, 495–500.

[21] a) I. E. Paulauskas, G. E. Jellison, L. A. Boatner, G. M. Brown, International Journal of Electrochemistry 2011, 2011, 1–10. b) F. E. Osterloh, B. A. Parkinson, MRS Bull. 2011, 36, 17–22.

[22] F. E. Osterloh, Chem. Mater. 2008, 20, 35–54.

[23] R. Peng, C.-M. Wu, J. Baltrusaitis, N. M. Dimitrijevic, T. Rajh, R. T. Koodali, Chem. Commun. 2013, 49, 3221.

[24] B. A. Gregg, A. J. Nozik, J. Phys. Chem. 1993, 97, 13441–13443.

[25] A. Kumar, N. S. Lewis, Appl. Phys. Lett. 1990, 57, 2730.

[26] a) A. Kudo, Pure Appl. Chem. 2007, 79, 1917–1927. b) A. K. A. T. K. D. K. M. a. T. O. M. Shibata, Chem. Lett.,, 1987, 1017–1018. c) T. S. T. Kawai, Nature (London, U. K.), 1979, 282-283.

[27] C. XING, Y. ZHANG, W. YAN, L. GUO, International Journal of Hydrogen Energy 2006, 31, 2018–2024.

[28] W. W. a. C. H. H. R. Dingle, Physical Review Letters 1974, 827–830.

[29] Eric Drexler, "There is plenty of room at the bottom - speech by Feynman", can be found under http://metamodern.com/2009/12/29/theres-plenty-of-room-at-the-bottom%E2%80%9D-feynman-1959/, 2013.

[30] H. M. Chen, R.-S. Liu, J. Phys. Chem. C 2011, 115, 3513–3527.

[31] T. R. Kline, M. Tian, J. Wang, A. Sen, M. W. H. Chan, T. E. Mallouk, Inorg. Chem. 2006, 45, 7555–7565.

[32] S. Tschierlei, M. Karnahl, M. Presselt, B. Dietzek, J. Guthmuller, L. González, M. Schmitt, S. Rau, J. Popp, Angewandte Chemie 2010, 122, 4073–4076.

[33] T. Ikeda, A. Xiong, T. Yoshinaga, K. Maeda, K. Domen, T. Teranishi, J. Phys. Chem. C 2013, 117, 2467–2473.

[34] Y. Yuan, N. Yan, P. J. Dyson, ACS Catal. 2012, 2, 1057–1069.

[35] A. J. Nozik, Annu. Rev. Phys. Chem. 1978, 29, 189–222.

[36] M. A. Holmes, T. K. Townsend, F. E. Osterloh, Chem. Commun. 2011, 48, 371.

[37] R. A. Marcus 1964, 155–196.

[38] T. Yokoi, J. Sakuma, K. Maeda, K. Domen, T. Tatsumi, J. N. Kondo, Phys. Chem. Chem. Phys. 2011, 13, 2563.

[39] S. D. Tilley, M. Cornuz, K. Sivula, M. Grätzel, *Angewandte Chemie International Edition* 2010, *49*, 6405–6408.

[40] M. Gondal, A. Suwaiyan, *Applied Catalysis A: General* 2004, *268*, 159–167.

[41] T. K. Townsend, N. D. Browning, F. E. Osterloh, *ACS Nano* 2012, *6*, 7420–7426.

[42] J. B. González-Campos, E. Prokhorov, I. C. Sanchez, J. G. Luna-Bárcenas, A. Manzano-Ramírez, J. González-Hernández, Y. López-Castro, R. E. del Río, *Journal of Nanomaterials* 2012, *2012*, 1–11.

[43] V. Subramanian, E. E. Wolf, P. V. Kamat, *J. Am. Chem. Soc.* 2004, *126*, 4943–4950.

[44] K. Maeda, K. Domen, *J. Phys. Chem. C* 2007, *111*, 7851–7861.

[45] H. X. Dang, N. T. Hahn, H. S. Park, A. J. Bard, C. B. Mullins, *J. Phys. Chem. C* 2012, *116*, 19225–19232.

[46] J. Liebig, *Ann. Phys. Chem.* 1835, *110*, 570–613.

[47] Y. Wang, X. Wang, M. Antonietti, *Angew. Chem.* 2012, *124*, 70–92.

[48] Corkill, Cohen, *Phys. Rev., B Condens. Matter* 1993, *48*, 17622–17624.

[49] A. Thomas, A. Fischer, F. Goettmann, M. Antonietti, J.-O. Müller, R. Schlögl, J. M. Carlsson, *J. Mater. Chem.* 2008, *18*, 4893.

[50] Q. Guo, Y. Xie, X. Wang, S. Lv, T. Hou, X. Liu, *Chemical Physics Letters* 2003, *380*, 84–87.

[51] B. V. Lotsch, M. Döblinger, J. Sehnert, L. Seyfarth, J. Senker, O. Oeckler, W. Schnick, *Chem. Eur. J.* 2007, *13*, 4969–4980.

[52] F. Goettmann, A. Fischer, M. Antonietti, A. Thomas, *Angew. Chem.* 2006, *118*, 4579–4583.

[53] Bettina Lotsch, *Dissertation*, Ludwig-Maximilians-Universität München, München, 2006.

[54] Y. C. Zhao, D. L. Yu, H. W. Zhou, Y. J. Tian, O. Yanagisawa, *J Mater Sci* 2005, *40*, 2645–2647.

[55] A. Thomas, F. Goettmann, M. Antonietti, *Chem. Mater.* 2008, *20*, 738–755.

[56] E. Kroke, M. Schwarz, E. Horath-Bordon, P. Kroll, B. Noll, A. D. Norman, *New J. Chem.* 2002, *26*, 508–512.

[57] F. Goettmann, A. Fischer, M. Antonietti, A. Thomas, *Angew. Chem. Int. Ed.* 2006, *45*, 4467–4471.

[58]   T. Komatsu, *J. Mater. Chem.* 2001, *11*, 799–801.

[59]   E. G. Gillan, *Chem. Mater.* 2000, *12*, 3906–3912.

[60]   J. Zhang, X. Wang, Q. Su, L. Zhi, A. Thomas, X. Feng, D. S. Su, R. Schlögl, K. Müllen, *J. Am. Chem. Soc.* 2009, *131*, 11296–11297.

[61]   F. Goettmann, A. Fischer, M. Antonietti, A. Thomas, *Chem. Commun.* 2006, 4530.

[62]   X. Jin, V. V. Balasubramanian, S. T. Selvan, D. P. Sawant, M. A. Chari, G. Q. Lu, A. Vinu, *Angew. Chem.* 2009, *121*, 8024–8027.

[63]   J. Zhu, Y. Wei, W. Chen, Z. Zhao, A. Thomas, *Chem. Commun.* 2010, *46*, 6965.

[64]   a) Y. Wang, J. Yao, H. Li, D. Su, M. Antonietti, *J. Am. Chem. Soc.* 2011, *133*, 2362–2365. b) Y. Di, X. Wang, A. Thomas, M. Antonietti, *ChemCatChem* 2010, *2*, 834–838.

[65]   M. Deifallah, P. McMillan, F. Cora, *J. Phys. Chem. C* 2008, *112*, 5447–5453.

[66]   K. Maeda, K. Teramura, D. Lu, N. Saito, Y. Inoue, K. Domen, *Angew. Chem. Int. Ed.* 2006, *45*, 7806–7809.

[67]   B. Klahr, S. Gimenez, F. Fabregat-Santiago, J. Bisquert, T. W. Hamann, *J. Am. Chem. Soc.* 2012, *134*, 16693–16700.

[68]   M. Groenewolt, M. Antonietti, *Adv. Mater.* 2005, *17*, 1789–1792.

[69]   Y. Wang, J. Zhang, X. Wang, M. Antonietti, H. Li, *Angewandte Chemie* 2010, *122*, 3428–3431.

[70]   J. Zhang, X. Chen, K. Takanabe, K. Maeda, K. Domen, J. D. Epping, X. Fu, M. Antonietti, X. Wang, *Angewandte Chemie International Edition* 2010, *49*, 441–444.

[71]   G. Liu, P. Niu, C. Sun, S. C. Smith, Z. Chen, G. Q. Lu, H.-M. Cheng, *J. Am. Chem. Soc.* 2010, *132*, 11642–11648.

[72]   J. Zhang, J. Sun, K. Maeda, K. Domen, P. Liu, M. Antonietti, X. Fu, X. Wang, *Energy Environ. Sci.* 2011, *4*, 675.

[73]   Y. Zhang, A. Thomas, M. Antonietti, X. Wang, *J. Am. Chem. Soc.* 2009, *131*, 50–51.

[74]   A. Mishra, M. K. R. Fischer, P. Bäuerle, *Angew. Chem. Int. Ed.* 2009, *48*, 2474–2499.

[75] K. Takanabe, K. Kamata, X. Wang, M. Antonietti, J. Kubota, K. Domen, *Phys. Chem. Chem. Phys.* 2010, *12*, 13020.

[76] R. Bandari, W. Knolle, A. Prager-Duschke, H.-J. Gläsel, M. R. Buchmeiser, *Macromol. Chem. Phys.* 2007, *208*, 1428–1436.

[77] Y. Dong, K. He, L. Yin, A. Zhang, *Nanotechnology* 2007, *18*, 435602.

[78] A. Schäfer, H. Horn, R. Ahlrichs, *J. Chem. Phys.* 1992, *97*, 2571.

[79] F. Weigend, R. Ahlrichs, *Phys Chem Chem Phys* 2005, *7*, 3297–3305.

[80] S. Grimme, J. Antony, S. Ehrlich, H. Krieg, *J. Chem. Phys.* 2010, *132*, 154104.

[81] V. Filipe, A. Hawe, W. Jiskoot, *Pharm Res* 2010, *27*, 796–810.

[82] H. Shi, *J. Electrochem. Soc.* 1996, *143*, 3466.

[83] J. Zhang, M. Zhang, S. Lin, X. Fu, X. Wang, *Journal of Catalysis* 2013.

[84] D. Richter, C. Zschiesche, R. Glaeser, *Preprints of Symposia - American Chemical Society, Division of Fuel Chemistry, 2012,* 141–142.

[85] J. Liu, Y. Zhang, L. Lu, G. Wu, W. Chen, *Chem. Commun.* 2012, *48*, 8826.

[86] J. Zhang, M. Grzelczak, Y. Hou, K. Maeda, K. Domen, X. Fu, M. Antonietti, X. Wang, *Chem. Sci.* 2012, *3*, 443.

Appendix

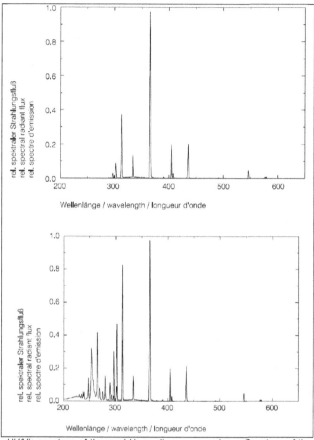

Figure 0-1: *UV/Vis spectrum of the used Hg medium pressure lamp. Spectrum of the lamp with an amorphous glass filter (upper figure) and without (lower figure)*

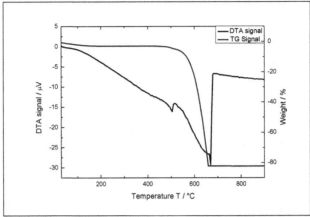

Figure 0-2: *TG and DTA from melamine derived g-C₃N₄*

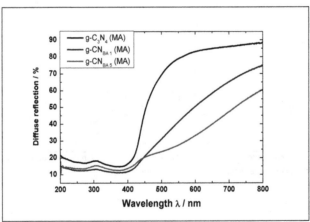

Figure 0-3: *Exemplary UV/Vis spectra from melamine derived carbon nitrides with different amounts of copolymer*

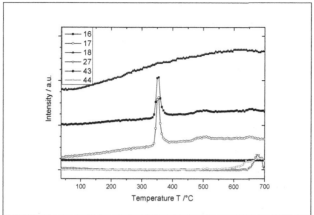

Figure 0-4: *Time depending mass spectra of the degassing products during polymerization process of melamine. Box-numbers refer to m/z ratios of the degassing products.*

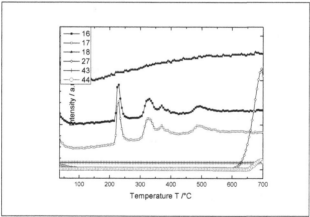

Figure 0-5: *Time depending mass spectra of the degassing products during polymerization process of melamine. Box-numbers refer to m/z ratios of the degassing products.*

Table 0-1: Results from N_2-Sorption for different g-C_3N_4

Compound	A_{BET} / m^2 g^{-1}	D_P / mm^3 g^{-1}
g-C_3N_4 (MA)	13.6	0.6
g-$CN_{BA\ 1}$ (MA)	10.1	-
g-$CN_{BA\ 5}$ (MA)	9.0	0.9
Activated g-C_3N_4 (MA)	39.1	0.3
g-C_3N_4 (DCA)	24.7	-
g-$CN_{BA\ 1}$ (DCA)	11.5	3.3
g-$CN_{BA\ 5}$ (DCA)	14.3	-
Activated g-C_3N_4 (DCA)	22.6	2.3

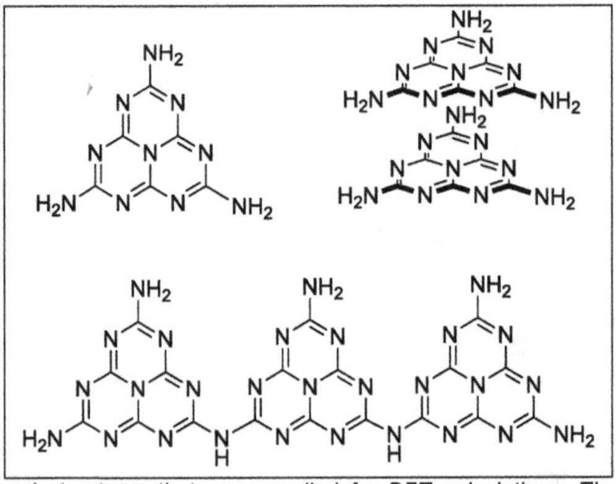

Figure 0-6: Additional structures that were applied for DFT calculations. The upper left structure shows a single heptazine, while the upper right structure describes two stacked heptazine rings. Additionally, DFT calulations were carried out for a polymer one-dimensional system of three heptazines (lower structure).

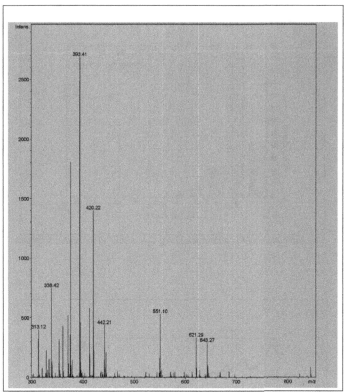

Figure 0-7: *Maldi mass spectra from Dye$_{Kov}$@g-C$_3$N$_4$ (covalent dye modification) derived from mela-mine*

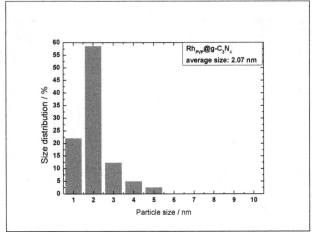

Figure 0-8: Size distributions of Pt nanoparticles on g-C₃N₄ from PVP mediated synthesis

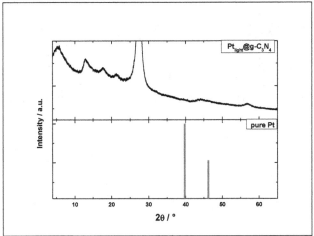

Figure 0-9: XRD patterns from Pt_light@g-C₃N₄

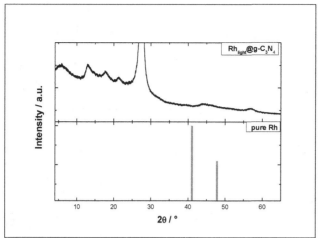

Figure 0-10: *XRD patterns from Rh$_{light}$@g-C$_3$N$_4$*

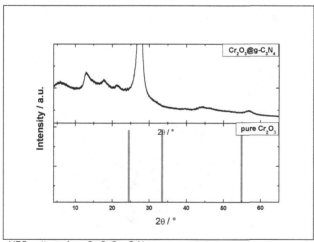

Figure 0-11: *XRD patterns from Cr$_2$O$_3$@g-C$_3$N$_4$*

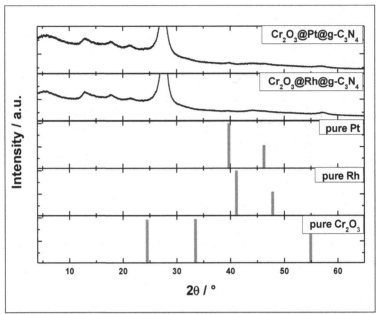

Figure 0-12: XRD patterns from $Cr_2O_3@NM_{PVP}@g\text{-}C_3N_4$

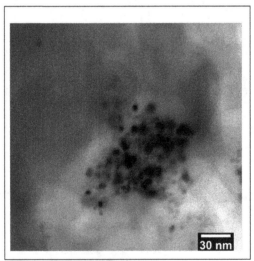

Figure 0-13: TEM images from $Cr_2O_3@Pt_{PVP}@g\text{-}C_3N_4$

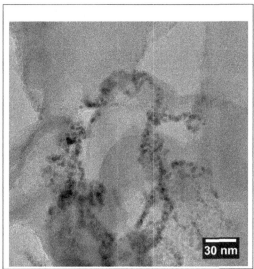

Figure 0-14: TEM images from $Cr_2O_3@Rh_{PVP}@g\text{-}C_3N_4$

Printed in the United States
By Bookmasters